T0100391

THE RISE OF
CHATGPT

www.royalcollins.com

THE RISE OF
CHATGPT

An Insight into the Next Era of AI

Kevin Chen

Books Beyond Boundaries

ROYAL COLLINS

The Rise of ChatGPT
An Insight into the Next Era of AI

Kevin Chen

First published in 2023 by Royal Collins Publishing Group Inc.
Groupe Publication Royal Collins Inc.
550-555 boul. René-Lévesque O Montréal (Québec) H2Z1B1 Canada

ISBN: 978-1-4878-1159-4

To find out more about our publications, please visit www.royalcollins.com.

Contents

Preface   *ix*

1

ChatGPT: An Explosive Arrival   1

1.1 The Emergence of ChatGPT   1
1.2 Reasons behind ChatGPT's Power   6
1.3 The "ChatGPT +" Effect   10
1.4 The Popularity of AIGC   12

2

AGI: Approaching the Singularity   17

2.1 Human Intelligence and AI   17
2.2 From ANI to AGI   23
2.3 ChatGPT Revolutionizing AGI Potential   27
2.4 More Powerful Versions of ChatGPT   30
2.5 The Foreseeable Future of ChatGPT   37

3

Commercial Battles of ChatGPT   **41**

3.1 ChatGPT's Rise: Transforming OpenAI's Finances and Valuation   41

3.2 Microsoft's Response   51

3.3 Google's Response   55

3.4 Apple's Response   61

3.5 Meta's Challenges in Exploring AI   63

3.6 Amazon's Response   66

3.7 NVIDIA's Achievements   71

3.8 Challenges to Musk   77

4

Looking for China's ChatGPT   **83**

4.1 The Commercialization Fantasy of ChatGPT   83

4.2 Baidu: Sprinting to Launch the Chinese Version of ChatGPT   86

4.3 Alibaba: Building Specialized ChatGPT-like Products   92

4.4 Tencent: Bullish on AIGC and Bolstering AIGC   97

4.5 ByteDance: Embracing the Wave and Challenges of ChatGPT   100

4.6 JD Cloud: Building an Industry Version of ChatGPT   104

4.7 ChatGPT: Reshaping the AI Industry   107

4.8 China-US Divide in ChatGPT   113

5

The Impacts of ChatGPT   **121**

5.1 ChatGPT's Impact on Search Engines   121

5.2 ChatGPT's Impact on Content Production   126

5.3 ChatGPT's Impact on Medical Field   132

5.4 ChatGPT's Impact on Legal Field   140

5.5 ChatGPT's Impact on Education Field   144

5.6 ChatGPT's Impact on New Retail   148

5.7 ChatGPT's Impact on Financial Industry   152

5.8 ChatGPT's Impact on Humanoid Robots   156

6

Are Humans Ready?   **161**

6.1 The Imperfect ChatGPT   161

6.2 The Realities of ChatGPT's Dilemmas   165

6.3 The Copyright Controversy of ChatGPT   169

6.4 ChatGPT's Impact on Employment   175

6.5 ChatGPT's Direction   186

Epilogue   *193*

Bibliography   *197*

Index   *199*

Preface

Recently, ChatGPT has taken the Internet by storm.

On November 30, 2022, OpenAI released a new model, ChatGPT. Thanks to its astonishing capabilities, within just five days of its launch it attracted one million users. Only two months later, ChatGPT's monthly active users (MAU) reached 100 million, making it the fastest-growing consumer application in history. By comparison, according to Sensor Tower data, it took the international version of TikTok about nine months to reach 100 million MAU. In contrast, Instagram took two and a half years.

The explosive growth of ChatGPT is rooted in its unprecedented capabilities. It has a mature and even astonishing understanding and creative ability. Besides writing code, scripts, and composing lyrics, ChatGPT engages in smooth conversations, fully demonstrating its dialectical analysis skills. ChatGPT even dares to question incorrect premises and assumptions, actively admits mistakes, and refuses unreasonable questions. The key reason for its success is that before ChatGPT, many artificial intelligence (AI) products were merely superficial, only engaging in simple data statistics and classification work. However, ChatGPT has provided AI with human-like language logic and communication capabilities, becoming the AI everyone anticipated.

More importantly, ChatGPT's success proves the large model technology route. This means that AI has finally moved from the previous big data statistics classification stage to today's human-like logic communication stage, and AI's evolution speed will surpass our expectations due to its powerful learning capabilities.

Based on the large model technology route, ChatGPT is like a universal task assistant, capable of integrating with different industries and giving rise to many application scenarios. In a sense, ChatGPT has opened the door to general AI, truly grounding AI. Elon Musk lamented, "We're not far from AI that is powerful enough to be dangerous," while Bill Gates said that the importance of chatbot ChatGPT is no less significant than the invention of the Internet.

ChatGPT's overnight fame has also quickly triggered a shock wave globally, setting off a boom in the Chinese and American AI industries, with AI companies fully entering the field and shaking the capital markets. First, ChatGPT's monetization plan surfaced, charging $20 monthly for the paid subscription version of ChatGPT Plus. Then, Microsoft expanded its partnership with ChatGPT's parent company, OpenAI, including gradually implementing $10 billion in new investments while integrating all its products with ChatGPT, among other initiatives. Additionally, on February 8, Beijing time, Google announced the launch of an experimental AI service called Bard.

In China, as there is no "Chinese version of ChatGPT" yet, Chinese Internet tech giants have embarked on the journey to find their version of ChatGPT. On February 7, Baidu officially announced the launch of a ChatGPT-like application and a large-scale natural language processing project called "ERNIE Bot," which will complete internal testing in March and be open to the public. On the evening of February 8, Alibaba, the Internet giant with a market capitalization of ¥1.89 trillion, confirmed that the company was developing its own chatbot, Tongyi Qianwen, which is currently in the internal testing phase.

In addition to causing tremors in the commercial capital markets, ChatGPT is also impacting humanity itself, raising questions about whether it can replace humans and the ethical issues it brings. In fact, the emergence of any new technology, especially revolutionary ones, will spark debate. Objectively speaking, the era of AI is an inevitable trend, and ChatGPT has just brought us closer to the AI era we envision. While many of us were unprepared to embrace it, ChatGPT arrived suddenly, was capable of genuinely helping us with our work and even outperforming humans in some tasks.

ChatGPT represents the qualitative change in AI, heralding a new AI revolution. This book is based on that premise, focusing on the birth and explosion of ChatGPT and the technology route behind its success. It provides a detailed analysis of the commercial impact of ChatGPT, involving globally recognized Internet tech giants such as OpenAI, Microsoft, Google, Baidu, Tencent, and Alibaba. The shock waves ChatGPT has created in the capital markets further illustrates its transformative power. This book also explores the social impact of ChatGPT, as its emergence signifies the beginning of a true AI era, with human-machine collaboration rapidly approaching.

The text is written in plain language, easy to understand, and engaging, with in-depth yet accessible content that progresses gradually. It helps readers comprehend the suddenly emergence of ChatGPT and sifts through the complex information to gain insights into the AI industry's transformation and the upcoming era of general AI.

AI is not only a technological marker of our time, but its technology-driven changes are also shaping this era. We need to be prepared for what is coming.

1

ChatGPT: An Explosive Arrival

1.1 The Emergence of ChatGPT

From the end of 2022 to the beginning of 2023, ChatGPT created by OpenAI spread like wildfire across the Internet, becoming a phenomenal application in the field of AI. Due to its astonishing capabilities, ChatGPT gained over a million registered users within five days of its release. Facebook took ten months to achieve the same milestone. According to a report by UBS, ChatGPT's MAU exceeded 100 million by the end of January 2023, making it the fastest-growing consumer application in history.

So, what exactly is ChatGPT, and how did it suddenly become so popular?

1.1.1 The Omnipotent ChatGPT

ChatGPT is the latest AI language model released by OpenAI, representing a significant breakthrough in natural language processing (NLP). Unlike previous intelligent voice assistants, this AI language model is surprisingly smart. Compared to current AI customer service systems, ChatGPT has genuinely evolved from artificial stupidity to AI, taking on the form we had long anticipated. Many describe it as being omnipotent—capable of chatting, searching, translating, writing poetry, academic papers, code, developing mini-games, participating in American college entrance exams, and conducting research and medical diagnosis. Foreign media have commented that ChatGPT could become the next disruptive technology in the industry.

Generative Pretrained Transformer (GPT) is a text-generating deep learning model trained on available Internet data. ChatGPT is derived from GPT-3, released by OpenAI in 2020, and is the largest AI model regarding training parameters. When GPT-3 was first released, it also caused a similar sensation. GPT-3 demonstrated various capabilities at that time, including answering questions, translating, writing articles, and performing mathematical calculations and code. Articles written by GPT-3 were almost indistinguishable from those written by humans. In OpenAI's tests, human evaluators struggled to determine the authenticity of the news, with an accuracy rate of only 12%. GPT-3 was considered the most potent language model then, and some netizens even described GPT-3 as "omnipotent."

Now, ChatGPT seems to be even more powerful than GPT-3. It engages in long creative conversations, answers questions, and writes various material based on user requirements, such as business plans, promotional material, poetry, jokes, computer code, and movie scripts. In short, ChatGPT possesses human-like logic, thinking, and communication capabilities, and its communication abilities in some fields are astonishing, surpassing expert-level conversations.

ChatGPT can also engage in literary creation. For example, it can write a novel framework if given a topic. When we asked ChatGPT to write a novel framework based on the theme of "AI changing the world," ChatGPT clearly provided the story background, protagonist, plot, and ending. After refining our request, ChatGPT continued to answer and complete the framework. ChatGPT has developed some memory capabilities, allowing for a continuous dialogue. Some users have praised "ChatGPT's language organization, text level, and logical abilities" as "stunning." Some even plan to delegate daily reports, weekly reports, and reflective summaries to ChatGPT for assistance.

Ordinary text creation is just the beginning. ChatGPT also helps programmers find bugs in their code. Some developers have said that ChatGPT provided highly detailed solutions to their technical issues, which is more reliable than some search software. Amjad Masad, the CEO of the American code hosting platform Replit, tweeted that ChatGPT is an excellent "debugging partner" because "it not only explains errors but also fixes them and explains the repair methods."

Regarding business logic, ChatGPT is highly knowledgeable about its strengths and weaknesses. It can perform competitive analysis, write marketing reports, and even has a solid grasp of the world economic situation, offering its insights.

ChatGPT also dares to challenge incorrect premises and assumptions, admitting mistakes and acknowledging questions it cannot answer. It actively refuses unreasonable queries, improving its understanding of user intent and increasing the accuracy of its responses. Some users have already attempted to have ChatGPT take the American college entrance exams; trick it into planning the destruction of the world; or even have ChatGPT impersonate OpenAI, creating a "ChatGPT within ChatGPT" scenario.

The rapid growth and diverse capabilities of ChatGPT have undoubtedly attracted widespread attention and admiration. Its ability to assist in various fields, from writing to debugging, demonstrates its potential to revolutionize how we work, learn, and communicate. The future of AI language models like ChatGPT holds enormous potential, and it will be fascinating to see how they continue to develop and impact our world.

1.1.2 Imperfections of ChatGPT

Although ChatGPT's performance has improved compared to the GPT-3 model, it is still imperfect. In fact, the human-like outputs and astonishing versatility of ChatGPT and GPT-3 are the results of excellent technology, not genuine intelligence. Both models can make ridiculous mistakes, especially in cultural common sense and mathematical calculations. Moreover, ChatGPT's responses are often lengthy and verbose, appearing logically consistent but sometimes just "deceiving" people. This is an inevitable drawback of this approach because it essentially generates data through probability maximization rather than logical reasoning.

While this kind of fabrication can be useful in some areas, such as for game developers, science fiction authors, and artists who use AI for inspiration, it is a significant disadvantage in scenarios that require accurate answers to specific questions. If non-experts cannot discern the accuracy of ChatGPT's answers, they can be severely misled. We can imagine the cognitive harm that a giant machine with near-zero content creation cost, 80% accuracy, and the ability to confuse non-experts 100% of the time would have if it took over the writing of all encyclopedias and answered all questions on platforms like Quora.

As a result, ChatGPT has been banned by various organizations. Stack Overflow recently banned ChatGPT, stating that the temporary ban was due to the low accuracy of its generated answers, which were harmful to the site and users seeking correct information. Additionally, top AI conferences have begun prohibiting ChatGPT and AI tools for writing academic papers. The International Conference on Machine Learning (ICML) believes that although language models like ChatGPT represent a future trend, they also bring unexpected consequences and difficult-to-solve problems. ICML states that ChatGPT is trained on public data, often collected without consent, making it difficult to hold anyone accountable for issues.

Apart from providing less accurate results, ChatGPT cannot cite information sources and needs to be made aware of events after 2021. Although its output is often smooth enough to pass muster in high school or college classrooms, it needs to achieve the careful wording of human experts.

People seem to have low standards for an intelligence. If something appears intelligent, we are easily deceived into thinking it is intelligent. The fact is that AI's greatest trick is making the world believe it exists. ChatGPT and GPT-3 represent a significant leap in this area, but they remain tools created by humans.

ChatGPT is based on data from 2021 and earlier and has limited usage, so some knowledge gaps and conversational mishaps are expected. However, with extensive user conversation training and massive data updates, ChatGPT will evolve at a pace beyond our imagination.

1.1.3 The Decisive Technologies of 2023

Omnipotent or not, perfect or not, as a phenomenal application in the field of AI, ChatGPT has taken the stage in history. It has begun to enter and even influence people's lives. From Silicon Valley tech giants to primary and secondary capital markets, everyone interested in the topic is discussing the future development of ChatGPT and the impact of AI technology.

In fact, when ChatGPT first launched, it was mainly popular within the AI and technology circles. After the Chinese New Year in 2023, its popularity continued to heat up. After February 2023, significant news regarding ChatGPT increased noticeably. People found that ChatGPT could easily write copy, code, cover various fields such as history, culture, and technology, and even write poetry, seek medical advice, fix bugs, write code, write papers, and compose lyrics. ChatGPT even passed a Google coding engineer interview with an annual salary of $183,000, seemingly capable of anything. The Internet was flooded with information about ChatGPT.

A report published by UBS Group showed that ChatGPT had an average of about 13 million unique visitors per day in January 2023, twice the number in December 2022. By the end of January 2023, ChatGPT's MAU exceeded 100 million. ChatGPT set a new record for user growth—in comparison, it took Instagram two and a half years to reach 100 million users.

On February 2, Microsoft announced that all its products would fully integrate ChatGPT. According to related news, it is expected that ChatGPT will be built into Bing search in March; Baidu will launch a generative search based on ChatGPT in March; the British journal *Nature* will no longer support AI-generated authorship; digital media company BuzzFeed plans to use OpenAI's AI technology to assist in creating personalized content; the University of Pennsylvania claims that ChatGPT can pass the final exams for the school's MBA program; OpenAI announced the development of an AI Text Classifier tool to help users distinguish whether the text is generated by ChatGPT or not.

At the same time, from a capital market perspective, ChatGPT's popularity has driven the growth of AI-related company stocks. In the first week after Spring Festival, ChatGPT and AI-generated content (AIGC) concept stocks were active, and related individual stocks continued to rise. Wind data shows that on February 3, the ChatGPT index continued to surge 5.56%, with a weekly increase of 30.18%. Leading concept stocks included companies like Seewo Intelligence, Haitian Rui Voice, CloudWalk Technology, Chuling Information, and Hanwang Technology, with weekly stock price increases of up to 60% and 70%. For example, Hanwang Technology, despite its forecasted net profit for 2022 at −¥140 to −¥98 million, still experienced continuous price increases thanks to the ChatGPT concept, gaining five consecutive price increases after the holiday.

Several listed companies have actively responded to investors regarding their layouts in related fields. For example, Jiecheng Holdings stated that its invested subsidiary Shiyou Technology's digital human has already accessed ChatGPT and is training a personalized model based on OpenAI using the digital human's character background and related datasets. Baidu also announced that it would release a ChatGPT-like product in March, and Alibaba's Damo Academy announced it is also developing a ChatGPT-like product.

Based on calculations with 100 million users and a monthly fee of $20, ChatGPT's annual revenue will exceed $20 billion. At the same time, more than one billion potential users worldwide can benefit from ChatGPT. The entire ChatGPT market size is estimated to exceed $200 billion. If ChatGPT's business model is successful, it will present huge profit prospects for investors.

Nowadays, there are numerous companies associated with ChatGPT-related concepts. According to CB Insights, there are currently about 250 startups in the ChatGPT concept field, with 51% in the A-round or angel round financing stage. In 2022, the ChatGPT and AIGC fields attracted more than $2.6 billion in investments and produced six unicorn enterprises,* with the highest-valued one being OpenAI, valued at $29 billion.

As Bill Gates has expressed, the rise of AI like ChatGPT will be as important as the birth of the Internet or the development of personal computers. The last time the AI industry was this bustling was when AlphaGo defeated Lee Sedol. ChatGPT has only been around for two months but has already significantly impacted people's lives, production, and businesses. This impact is different from the concept hype brought about by the emergence of the metaverse; it is a real transformation of human society's production and way of life. More accurately, ChatGPT has caused such a sensation in the technology field because it represents a breakthrough and application of AI technology genuinely advancing towards human-like intelligence.

1.2 Reasons behind ChatGPT's Power

It seems that ChatGPT is powerful and intelligent, capable of creating content and writing code. Its abilities in multiple aspects far exceed people's expectations. So, how does ChatGPT become so strong? Where do its various powerful abilities come from?

---

*"Unicorn Enterprise" refers specifically to a company that has been established for no more than ten years, has a valuation exceeding $1 billion, and has received private funding but has not yet gone public.

1.2.1 An Outstanding NLP Model

Behind its robust features, the technology is not mysterious. Essentially, ChatGPT is an outstanding new NLP model. Today, when most people hear about NLP, they first think of voice assistants like Alexa and Siri. NLP's basic function is to enable machines to understand human input, but this is just the tip of the iceberg. NLP is a subset of AI and machine learning (ML) focused on enabling computers to process and understand human language. While speech is part of language processing, the most critical advances in NLP lie in its ability to analyze written text.

ChatGPT is a pretrained language model based on the Transformer model. It learns natural language knowledge and grammatical rules by training on a massive text corpus. When users ask, it generates answers by analyzing and understanding the questions. The Transformer model provides a parallel computation method, allowing ChatGPT to generate answers quickly.

What is the Transformer model? To understand this, we need to look back at the development of NLP technology. Before the Transformer model, the mainstream model in NLP was the Recurrent Neural Network (RNN) with the addition of the Attention mechanism. RNN models are suitable for processing sequential data, such as language, and the Attention mechanism allows AI to understand the context. However, RNN + Attention models slow the entire model's processing speed because RNN processes words individually. There are problems with model instability or premature stopping of effective training when handling longer sequences, such as long articles or books.

In 2017, the Google Brain team published a paper at the Neural Information Processing Systems conference called "Attention is all you need." The authors proposed the Transformer model based on self-attention mechanisms for the first time and applied it to NLP. Compared to the previous RNN models, the 2017 Transformer model could simultaneously perform data calculation and model training in parallel, had shorter training times, and produced models with interpretable syntax.

The initial Transformer model had 65 million adjustable parameters. The Google Brain team trained the initial Transformer model using various public language datasets. These datasets included the 2014 English-German Machine Translation Workshop (WMT) dataset (with 4.5 million English-German corresponding sentence pairs), the 2014 English-French Machine Translation Workshop dataset (36 million English-

French corresponding sentence pairs), and parts of the University of Pennsylvania Treebank language dataset (40,000 sentences from the *Wall Street Journal* and an additional 17 million sentences from the Treebank). Moreover, the Google Brain team provided the model's architecture in the paper, allowing anyone to build and train a similar model structure using their data.

After training, the initial Transformer model achieved industry-leading scores in various benchmarks, including translation accuracy and English constituent syntactic analysis, making it the most advanced large-scale language model at the time. ChatGPT also utilizes the technology and ideas of the Transformer model, extending and improving it to better suit language generation tasks. Based on the Transformer model, ChatGPT has achieved its success today.

1.2.2 ChatGPT's Data Training

Of course, even a great language model would only be useful with data. So, based on the Transformer model, ChatGPT's developers embarked on a massive data training process.

Even before ChatGPT, the developers had released GPT-1, GPT-2, and GPT-3. Although these earlier models didn't make as much noise, they were still quite large.

GPT-1 had 117 million parameters, and the developers used a classic large-scale book text dataset for pretraining the model. This dataset contained over 7,000 unpublished books, covering genres like adventure, fantasy, and romance. After pretraining, the model was further trained on four different language scenarios with specific datasets. The resulting model achieved better results than the base Transformer model in question answering, text similarity assessment, semantic implication judgment, and text classification, becoming the new industry leader.

In 2019, the company unveiled GPT-2, a model with 1.5 billion parameters. The model's architecture was similar to GPT-1, with the main difference being its larger scale. As expected, GPT-2 broke records for large language models (LLMs) in multiple language scenarios.

GPT-3's neural network astonishingly consisted of 175 billion parameters. Apart from being two orders of magnitude larger than GPT-2, the architecture had no fundamental difference. However, GPT-3 could generate complete, coherent long articles based on simple prompts, making it almost unbelievable that a machine created them. GPT-3 could also write program code, create recipes, and perform almost all text creation tasks. GPT-4 further improved in dialogue logic and more complex logic like mathematics, making AI and somewhat emotionally intelligent.

From GPT-1 to GPT-2, GPT-3, and the latest GPT-4, although ChatGPT's specific data has not been disclosed, it is almost certain that it used even more data for training. With breakthroughs in computing power and open ecosystem interfaces, it's only a matter of time before GPT-5 and GPT-6 surpass humans in knowledge-based domains, especially when integrating cameras, voice, and video functionality.

1.2.3 The Culmination of Advantages

It's worth noting that ChatGPT differs from GPT-3. In March 2022, OpenAI, ChatGPT's developer, published a paper titled "Training Language Models to Follow Instructions with Human Feedback" and introduced the InstructGPT model based on GPT-3 and further fine-tuned. During InstructGPT's training, human evaluation and feedback data were added, not just using pre-prepared datasets. In other words, unlike GPT-3, which learned from massive data, human feedback on results became part of the AI learning process in ChatGPT.

During GPT-3's public testing, users provided a large amount of dialogue and prompt data, while OpenAI's internal data labeling team also generated many artificially labeled datasets. This labeled data helped the model learn human labels while directly learning from the data. OpenAI then used this data to fine-tune GPT-3's supervised training.

Subsequently, they collected samples of answers generated by the fine-tuned model. Generally, the model gives countless answers for each prompt, but people usually want to see only one. The model needs to rank these answers and select the best one. The data labeling team manually scored and ranked all possible answers, choosing the answer that best suited human thinking and communication habits. These human-scored results could further establish a reward model, automatically providing feedback to the language model, encouraging it to produce better answers and suppressing undesirable ones. This reward model helps the AI find the optimal answer.

Finally, the team used the reward model and more labeled data to optimize the fine-tuned language model, iterating the process until they obtained the final model, InstructGPT.

In simpler terms, OpenAI's GPT-3, released in 2020, allowed computers to mimic human-like conversation for the first time. However, while GPT-3 could generate all sorts of responses, it didn't always consider the context, often resulting in incorrect or confusing answers. To address this, OpenAI introduced human supervisors to "teach" AI how to answer human questions better. When AI responses aligned with human evaluation standards, high scores were given, while low scores were given when they didn't meet those standards. This allowed AI to optimize data and parameters based on human values.

By harnessing all these advantages, ChatGPT demonstrated unprecedented capabilities, becoming a phenomenal application in the AI field.

1.3 The "ChatGPT +" Effect

Within two months of its release, ChatGPT attracted significant attention due to its impressive performance in generating human-like text. The success of ChatGPT led to people experiencing the convenience brought by AI, and soon, the "ChatGPT +" effect emerged.

1.3.1 The Meaning of the "ChatGPT +" Effect

The so-called "ChatGPT +" effect combines ChatGPT and other AI programs. One example is the integration of Wolfram Alpha and ChatGPT. On January 9, 2023, Wolfram posted an article on their website, comparing the highly popular ChatGPT with the 14-year-old Wolfram Alpha question-answering system to combine the two.

While ChatGPT demonstrated astounding abilities in creating text, its math capabilities could have been stronger, sometimes even making mistakes with simple arithmetic or elementary school problems. Wolfram Alpha, on the other hand, excels in answering questions related to science, technology, engineering, and mathematics. The combination of ChatGPT and Wolfram Alpha addressed ChatGPT's shortcomings.

Wolfram Alpha, developed by Stephen Wolfram, the creator of the Wolfram Language, is based on the idea that the world is computable. His goal was for computers to execute any task that users could describe. People define the objective, and computers try to understand and execute the meaning to the best of their abilities. Stephen Wolfram created the Wolfram Language and the computational knowledge search engine Wolfram Alpha to achieve this.

Wolfram Alpha was officially launched on May 18, 2009. Its backend computations and data processing are implemented using Mathematica. With Mathematica's support for geometric, numerical, and symbolic computation and its powerful visualization capabilities, Wolfram Alpha can answer a wide range of mathematical questions and present the results in a clear and visually appealing manner. This computational knowledge engine laid a solid foundation for Apple's digital assistant, Siri.

Stephen Wolfram believes his tool can perfectly complement ChatGPT, as Wolfram Alpha has powerful structured computing capabilities and can understand natural language.

For example, let's ask ChatGPT the distance between Chicago and Tokyo. It might not provide an accurate answer because its response is based on the specific distance it learned during training, which might need to be corrected. Even if the answer is correct, more than relying on such a simple solution is needed; a practical algorithm is needed. Wolfram Alpha, however, can efficiently use its structured, high-precision knowledge to convert a problem into an accurate calculation.

ChatGPT and Wolfram Alpha combine to create "ChatGPT+," providing an opportunity for ChatGPT to grow stronger and more versatile.

1.3.2 Applications of the "ChatGPT+" Effect

The emergence of the "ChatGPT+" effect has provided a reference for companies exploring the commercialization of AIGC. Some users have combined ChatGPT with Stable Diffusion (an AI text-to-image tool) by first generating random artistic prompts using ChatGPT and then using these prompts as input for Stable Diffusion, resulting in visually stunning artworks. Another combination is "ChatGPT+WebGPT," where WebGPT is a version of GPT developed by OpenAI that answers questions by querying search engines and summarizing the information found. "ChatGPT+WebGPT" provides real-time updates and more accurate judgments on the truthfulness of the information.

Microsoft CEO Satya Nadella revealed plans to integrate AI tools like ChatGPT and DALL·E into Microsoft products, including the Bing search engine, Office suite, Azure cloud services, and Teams chat software. Integrating ChatGPT into Bing can give users more comprehensive information and source links. At the same time, more powerful NLP systems can recognize keywords, providing more accurate and personalized content recommendations. In the Office suite, NLP technology will allow users to search content more flexibly and intelligently and help them quickly generate personalized text, bringing an intelligent upgrade to the office experience. ChatGPT is expected to grow rapidly in the office ecosystem, accelerating the commercialization of conversational AI and AIGC.

It is foreseeable that "ChatGPT +" will bring new gameplay and experiences to existing products and services, and AI applications will enter a new stage.

1.4 The Popularity of AIGC

2022 was a year when AIGC became incredibly popular, from AI-generated paintings to code and even literary works. People were amazed by AIGC, as it was on par with human levels of creativity. The late 2022 arrival of ChatGPT took AIGC to new heights. The influential journal *Science* listed AIGC as a breakthrough in the AI field in its top ten scientific breakthroughs of 2022. *Gartner* also ranked AIGC as one of the top five influential technologies of 2022. *MIT Technology Review* listed AI synthetic data as one of the top ten breakthrough technologies of 2022 and even referred to AIGC as the most promising progress in the AI field in the past decade.

1.4.1 AIGC Goes Viral

AIGC, a combination of AI and GC (generated content), refers to using AI to produce content. In the past, we've heard of PGC (professionally generated content) and UGC (user-generated content) as the main content creation methods. AIGC is quickly becoming the new content creation method following PGC and UGC. While PGC and

UGC involve humans as the primary content creators, AIGC shifts the content creator role to AI.

The concept of AIGC began in 2022. Previously, AI programs like Microsoft Xiaoice, which created poetry, writing, and songs, fell within the realm of AIGC. However, it wasn't until 2022, when an AI-generated painting went viral, that AIGC exploded. In August 2022, an AI-generated painting called *Space Opera House*, created by the AI tool Midjourney, won first prize in a digital artist competition in Colorado. This painting triggered a global debate and trended on social media, garnering over 110 million views in a single day.

In October 2022, Stability AI secured about $100 million in funding, valuing the company at $1 billion and catapulting it into the ranks of "unicorns." Stability AI released the open-source model Stable Diffusion, which generates images based on user input text prompts. Text-to-Image (T2I) models such as Stable Diffusion, DALL·E 2, and Midjourney have ignited the AI art creation domain, marking the infiltration of AI into the artistic field.

As AI-generated images gained popularity, ChatGPT emerged, taking human-computer interaction to new heights by engaging in seamless conversations with users across various topics. Those who have experienced ChatGPT have been impressed by its powerful capabilities. It easily engages in conversations across diverse fields and understands a wide range of needs, from fixing code bugs and creating novels to suggesting development strategies for Twitter, questioning incorrect assumptions, and rejecting unreasonable requests.

Since 2022, AIGC has officially entered the fast track of development. Now, major tech companies worldwide are embracing AIGC, continuously releasing relevant technologies, platforms, and applications.

In conclusion, the rapid growth and adoption of AIGC, as demonstrated by ChatGPT and other AI tools, have signaled a new era in AI applications. As more companies integrate these technologies into their products and services, we can expect a greater variety of innovative applications and user experiences, further expanding the potential of AIGC and transforming the way we interact with and utilize AI.

1.4.2 Expansive Potentials of AIGC

Whether it's the viral AI-generated art or the astonishing ChatGPT, both fall under the concept of AIGC. AIGC has not only demonstrated its capabilities in image and text generation but also held great potential in fields such as short videos, animation, and music.

Of course, image generation is currently the most rapidly developing aspect of AIGC, with numerous real-world applications. Depending on the use case, image generation can be divided into image editing tools and end-to-end image generation. Image editing includes attribute and content editing. End-to-end image generation includes methods based on existing images, such as generating complete images from sketches, and multimodal transformations, like generating images from text. Notable products or algorithmic models include EditGAN, Deepfake, DALL·E, Midjourney, Stable Diffusion, and Wenxin Yige.

AI text generation is another early development within AIGC. Based on use cases, it can be divided into non-interactive and interactive text generation. Non-interactive text generation includes content continuation, summary/headline generation, text style transfer, whole text generation, and generating text descriptions from images. Interactive text generation includes chatbots and text-based interactive games. Notable products or algorithmic models include JasperAI, Copy.ai, Caiyun Xiaomeng, AI Dungeon, and ChatGPT.

AI video generation can be divided into video editing (e.g., quality restoration, special effects, face-swapping, etc.), automated video editing, and end-to-end video generation (e.g., generating videos from text). Google's text-to-video AI system Phenaki is a typical example. Although the quality of the generated videos is currently low, they can already span two minutes, covering multiple scenes and varying themes. Phenaki's website demonstrates that a 200-word prompt can generate a video about a futuristic sci-fi world. As AI's involvement in short video production increases, AIGC will account for more content on short video platforms alongside traditional UGC and PGC. Its influence should not be underestimated.

Some AI audio generation technologies have already matured and are applied in various consumer products. Audio generation can be divided into Text-to-Speech (TTS) and music generation. TTS includes applications such as voice assistants, audiobook production, and intelligent dubbing. Music generation can create specific compositions based on initial melodies, images, text descriptions, music genres, and emotions. Notable products or algorithmic models include DeepMusic, WaveNet, Deep Voice, and MusicAutoBot.

In addition, AI generation encompasses code generation, game creation, and 3D generation. AIGC has entered a new era, and it is expected that AIGC will continue to penetrate various industries, enriching application scenarios in gaming, animation, and media. According to *Gartner*, by 2025, AI-generated data will account for 10% of all data, while the analysis in "Generative AI: A Creative New World" suggests that AIGC has the potential to generate trillions of dollars in economic value.

1.4.3 A Revolutionary Change in Content Production

If AI recommendation algorithms are the powerful engine for content distribution, then AIGC is the powerful engine for data and content production.

In traditional creation, humans are often considered the authoritative spokesperson and owner of the inspiration. In fact, it is due to human's radical creativity, irrational originality, and illogical laziness—rather than rigid logic—that machines have struggled to mimic these human traits, keeping creative production exclusive to humans. However, with the emergence and development of AIGC, the human-centric nature of creative authorship is being challenged, and artistic creation is no longer exclusive to humans. Even in imitative creation, AI's ability to mimic the formal styles of art pieces challenges the idea that creative authorship is an exclusively human patent.

AIGC is evolving toward higher efficiency, quality, and lower costs. In some cases, it produces better results than human-made creations. Industries requiring human knowledge creation, ranging from social media to gaming, advertising to architecture, coding to graphic design, product design to law, and marketing to sales, may all be influenced and transformed by AIGC. Massive amounts of data required for the digital economy and AI development can also be generated and synthesized through AIGC technology, i.e., synthetic data.

Today, AIGC is sparking a revolution in content production. Amid a high demand for content, the changes in content production methods brought by AIGC are causing shifts in content consumption patterns. For example, AI-generated art can increase the efficiency of art material production and has found preliminary applications in gaming and digital collectibles.

A prime example of text-generating AIGC is the popular ChatGPT. Not only does ChatGPT fulfill basic conversational functions with humans and answer follow-up questions but it also can admit mistakes, question incorrect premises, and refuse inappropriate requests. Moreover, ChatGPT handles various styles and genres based on user feedback across various fields. Its code editing ability and performance in various common text output tasks greatly exceed expectations.

Although AIGC seems to be a broader concept, it hasn't gained as much popularity as ChatGPT due to differences between the two. AIGC, despite its wider scope, primarily focuses on the semantic understanding and generation of images, which differs from ChatGPT's human-like intelligence based on neural networks. In comparison, ChatGPT embodies the AI humans have long anticipated, with human-like communication abilities and the capacity to become a powerful assistant by integrating vast amounts of data.

ChatGPT has transformed the long-discussed and awaited AI from artificial stupidity into a true AI-like form. Regardless of whether its technology is the most advanced, its appearance matches people's expectations. As for the future, whether AIGC will encompass ChatGPT or ChatGPT will replace the concept of AIGC with faster iterations and commercial applications remains uncertain.

However, regardless of how these technologies will be defined in the future, ChatGPT will replace all regular and rule-based tasks in human society or even more advanced AIGC and some creative work will accelerate into a human-machine collaboration era.

2

AGI: Approaching the Singularity

2.1 Human Intelligence and AI

Before the emergence of silicon-based intelligence, humans, as carbon-based beings, dominated Earth and constructed human civilization and order. However, after the emergence of silicon-based intelligence, especially as it enters the third stage of superintelligence, the world will no longer be solely dominated by carbon-based intelligence. Instead, a new world will form where silicon-based and carbon-based intelligence coexists and collaborate, mutually integrating.

2.1.1 The Origin of Intelligence

The Big Bang occurred 13.8 billion years in the past, marking the beginning of all history and the creation story.

Earth was born 4.6 billion years ago. Six hundred million years later, the earliest life emerged in the primitive oceans, and living organisms began their complex and lengthy evolution from prokaryotes to eukaryotes.

During the Ediacaran period, 600 million years before the present, Earth witnessed the appearance of multicellular Ediacaran organisms, and primitive coelenterates floated in the Ediacaran seas.

Controlling their movements were specialized cells within their bodies called neurons. Unlike other cells that primarily formed various tissue structures with nearby cells, neurons extended long nerve fibers from their cell bodies, connecting with the nerve fibers of other neurons. Most of these nerve fibers were dendrites, responsible for receiving and transmitting information, while there was only one axon (which could branch) for outputting information. When dendrites receive information exceeding the excitation threshold, the entire neuron emits a brief but very noticeable action potential, similar to a light bulb turning on. This potential would spread throughout the neuron, including the nerve fiber endings far from the cell body, almost instantly. Afterward, the synaptic structures between the previous neuron's axon and the next neuron's dendrite would be activated by the electrical signal, releasing neurotransmitters to transmit information between the two neurons and having different excitatory or inhibitory effects depending on their type. With their structural features, these earliest neurons formed a network distributed throughout the coelenterates' bodies. This seemingly rudimentary neural network laid the foundation for the basic structure of all future nervous systems.

Starting 20 million years ago, some primates began spending more time living on the ground. Around 7 million years ago, in a location in Africa, the first bipedal "hominids" appeared.

Two million years ago, another hominid species appeared in East Africa, called "Homo habilis." These creatures could craft simple stone tools. Over the next 1.5 million years, intelligence began to develop in their brains, which were only about half the size of modern human brains. They improved their stone tools and even attempted to tame wild flames. As a result of natural selection and genetic mutations, their descendants' brain capacities grew larger, eventually leading to the emergence of Homo erectus.

According to paleontological research, Homo erectus had a body size similar to modern humans, and their brain capacity was not much different from ours. They crafted more sophisticated and complex stone tools compared to Homo habilis. Subsequently, some members of this species left Africa and, through generations of reproduction and migration, reached as far as present-day China. Finally, our species, Homo sapiens, emerged in East Africa approximately 250,000 years ago, commencing the unique evolution of human intelligence.

2.1.2 *Human Intelligence*

Approximately 200,000 years ago, the human brain underwent a leap in development. The association cortex, especially the frontal lobe, experienced rapid growth, leading to high-energy consumption—the human brain accounts for only about 2% of body weight but consumes 20% of the energy. As a result of these trade-offs, the brain had many neurons for the first time, allowing for deep abstraction and organization of various information. Since then, human intelligence has evolved through the transmission of wisdom and adaptation to the environment through cultural factors, breaking free from the shackles of natural evolution and transforming humans from isolated apes in East Africa to highly invasive species that spread across the globe.

The first manifestation of human intelligence was the transformation of physical forms. In prehistoric times, humans' transformation of matter was very simple, starting with the transformation of basic geometric shapes, such as shaping stones into sharp or blunt stone axes. Early humans used these tools to attack wild animals, sharpen wooden sticks, or dig plant roots, making them versatile tools.

During the Middle Stone Age, tools evolved into inlaid tools, where handles made of wood or bone were attached to stone axes. This development led to the creation of complex tools such as stone knives, stone spears, and stone chains, culminating in the invention of the bow and arrow. In the Neolithic Age, humans learned to drill holes in stone tools, inventing stone sickles, stone shovels, stone hoes, and stone mortars and pestles for processing grains. The transformation of simple geometric shapes not only changed the forms of materials according to human needs but also exercised and changed the human brain, allowing it to take the first step toward becoming intelligent entities.

The second manifestation of human intelligence was the transformation of energy. The primitive understanding of "fire" and its relationship to humans is a clear example. From the fear of wildfires caused by lightning strikes to using fire for cooking, warming, lighting, and repelling wild animals, the mastery of artificial fire-making marked the true utilization of fire as a natural force. Engels pointed out that "the production of fire by friction for the first-time enabled man to control a natural force and ultimately separated him from the animal kingdom."

The use of fire led early humans to learn to make pottery, and pottery technology greatly advanced ancient material technology and material processing techniques. It was the first time that human processing of materials went beyond merely changing the geometric shape of materials and began to alter their physical and chemical properties. Additionally, the development of pottery technology laid the foundation for the later emergence of metallurgy.

The third manifestation of human intelligence was the transformation of information. In the process of transforming material forms and energy, the creations and tools used, such as stone axes, fire-making tools, and pottery, internalized the relationships and information between humans and nature, as well as between humans. They serve as both material and spiritual means, and as information carriers. As a result, people's engagement in transforming material forms and energy is inevitably accompanied by the transformation of information.

The transformation of information led humans to create language, allowing them to share common needs and feelings during the process of material transformation and to internalize the relationships and information between people and nature and between people in labor processes and outcomes. This formed a kind of "consensus" and represented different consensus contents with specific syllables.

The emergence of language allowed humans to summarize and extract abstract, general concepts from concrete objects and to describe, communicate, and even learn these concepts accurately through language. In fact, the creation of language is an inevitable result of early human evolution and is closely linked to the development of brain functions and other human functions.

The Broca area in the left frontal cortex of the human brain is responsible for language production, while the Wernicke area in the back governs language reception. The corresponding right hemisphere areas receive signals from the left hemisphere through the corpus callosum to perform more advanced functions, such as appreciating music, art, and spatial orientation. With approximately 200 million nerve fibers, the corpus callosum transmits information between the two hemispheres.

As mentioned by Hideo Kojima's close friend and renowned science fiction writer, Project Itoh, in *Genocidal Organ*, language is essentially an organ in the brain. However, due to the emergence of this brain structure, human development has experienced explosive growth, and humans have transformed from isolated apes in East Africa to highly invasive species that have spread throughout the world.

Subsequently, the "imagined community" based on language emerged, and human social behavior transcended the instinctive tribal level of primates, developing larger and more complex trends. With the invention of writing, the earliest civilizations and city-states were finally born in the Mesopotamian region of Western Asia.

2.1.3 *From Human Intelligence to AI*

The origins and development of human intelligence in terms of material form, energy, and information have shaped the beginnings of human intelligence and laid the groundwork for human intelligence activities related to material transformation.

Throughout the thousands of years since the cognitive revolution, agricultural revolution, and industrial revolution, human activities have demonstrated that there are three fundamental categories of things humans seek to understand and transform: matter, energy, and information. To date, the main technologies humans have mastered are related to transforming these three elements and are based on the development of material, energy, and information technology.

As the technology in these three basic fields continues to advance, human intelligence activities regarding material transformation have progressed from single-element transformation to multi-element transformation. The manufacture and use of steam engines represent a composite transformation of matter and energy by human intelligence. The production and application of electronic computers exemplify the integrated transformation of matter, energy, and information. Today, research on AI can be understood as the transformation of matter, energy, information, and human intelligence combined.

In 1950, Alan Turing published a paper titled "Computing Machinery and Intelligence," which raised the question whether machines could think and laid the groundwork for the birth of AI. In 1957, the first ML project was launched, marking the birth of AI as a discipline. Inspired by neuron theories, artificial neural networks were proposed as an essential AI algorithm and have been continuously improved over the subsequent decades. Like natural neural networks in the human brain, artificial neural networks use virtual "neurons" as basic computational units and arrange them in functional layers like neurons in the brain's cortex. However, there are still many differences in connection patterns and working principles between the two, so they should not be simply equated.

After countless iterations and setbacks, the development of AI in the 21st century has entered a new phase. The latest generation of neural network algorithms has demonstrated impressive performance in learning tasks. The accuracy of various image and audio recognition software has been increasing, and the intelligence level of language processing programs has been growing daily.

As a result, the endless expansion of human intelligence is transforming and reshaping the entire natural world and creating a new form of intelligence: machine intelligence.

2.1.4 *The Essence of Intelligence*

From human intelligence to AI, what is the essence of intelligence?

We know that human intelligence is mainly related to the vast association cortex of the human brain. These areas are not directly involved in sensory and motor functions and are relatively smaller in the brains of most animals. In humans, the massive number of neurons in the association cortex is the building block for the human soul. Abilities in which the human brain far surpasses other animals, such as language, declarative memory, and working memory, are closely related to the association cortex. Our brains, however, are enclosed within our skulls throughout our lives and can only perceive external electrical and chemical signals.

That is to say, the essence of intelligence is a complex "algorithm" that can deduce, learn, and reconstruct the features of the external world through limited input signals. From this perspective, the abstract concept of "intelligence" is indeed quite close to what Descartes called "spirit." However, it must still be inscribed on a specific material substrate, such as the brain's cortex or integrated circuits.

This also implies that AI, as a form of intelligence, will theoretically be able to run the algorithm known as "self-consciousness" sooner or later. Although some believe that AI will never surpass the human brain because humans themselves do not know how the brain works, the fact is that the speed at which humans iterate AI algorithms is much faster than the speed at which DNA iterates its algorithms through natural selection. Therefore, AI does not need to understand how the human brain works to surpass human intelligence.

Human intelligence and AI are two sets of intelligence that coexist in today's world. In fact, the "thinking patterns" of AI are entirely different from those of humans. AI is in its infancy compared to human intelligence, which has slower basic component processing speeds, encodes large amounts of unmodifiable primal instincts, and has a limited capacity for self-shaping. Still, it has much greater potential for future development.

In fact, AI, including AlphaGo, has already proven that machines will surpass humans in tasks with defined goals. In 20 years, AI based on deep learning and its "descendants" will outperform humans in many tasks, but humans will still be better at many others.

In the future, the more likely scenario may be pursuing a symbiotic relationship between humans and AI rather than obsessing over the strength of human intelligence versus AI or whether AI will replace human intelligence as the protagonist of the world. Today, the emergence of ChatGPT allows people to feel the power of AI genuinely. ChatGPT is unlike any previous AI product, as it performs as well or even better than humans in most tasks. This perhaps demonstrates the notion that humans are not the only gold standard of intelligence.

2.2 From ANI to AGI

AI is a broad concept, so many different types or forms of AI exist. Based on different capabilities, we can classify AI into three categories: narrow AI (ANI), general AI (AGI), and super AI (ASI).

2.2.1 The Current AI

So far, most AI products we encounter are still ANI.

ANI is an AI programmed to perform a single task, whether checking the weather, playing chess, or analyzing raw data to write news reports. ANI is also known as weak AI. It is worth mentioning that although some AI can defeat world chess champions, like AlphaGo, that is the only thing they can do. If you ask AlphaGo to find a better way to store data on a hard drive, it will look at you blankly.

Our smartphones are like little ANI factories. When we use map applications for navigation, check the weather, talk to Siri, or engage in many other daily activities, we are using ANI.

A classic type of ANI we often use is our email spam filter, which has the intelligence to determine what is spam and what is not and can learn from our specific preferences to help filter out junk mail.

ANI also works behind our online shopping experiences. For example, when you search for a product on an e-commerce website and then see it "recommended for you" on another website, you might feel creeped out. Behind this is a network of ANI systems working together, informing each other who you are and what you like, and then using that information to decide what to show you. Some e-commerce platforms often display "people who bought this also bought ..." on their home page, which is another ANI system. It collects information from millions of customers' behaviors and combines that data to upsell to you, making you buy more stuff cleverly.

ANI is like the early-stage of computer development when people first designed electronic computers to replace human calculators for specific tasks. Mathematicians like Alan Turing believed that we should create general-purpose computers that can be programmed to perform any task.

So, there was a transitional period when people made various computers, including task-specific ones, analog computers, those that could only change their purpose by altering circuits, and some that worked on decimal rather than binary systems. Nowadays, almost all computers conform to the universal form envisioned by Turing, which we call the "Universal Turing Machine." With the right software, today's computers can perform almost any task.

Market forces determined that general-purpose computers were the right direction for development. Today, even though customized solutions like dedicated chips can perform specific tasks faster and more energy-efficiently, people still prefer low-cost, convenient general-purpose computers.

A similar shift is anticipated for AI today—people hope that AGI will emerge, which will be more similar to humans, capable of learning about almost anything, and able to perform multiple tasks.

2.2.2 AGI and ASI

Unlike ANI, which only performs a single task, AGI refers to AI technology that addresses multiple and even generalized problems without specifically encoding knowledge and application areas. Intuitively, ANI and AGI may seem like the same thing, only with a less mature and complex implementation, but they are different. AGI can reason, plan, solve problems, think abstractly, understand complex ideas, learn quickly, and learn from experience, just like humans.

Of course, AGI is not omniscient. Like any other intelligent being, it needs to learn different knowledge contents based on the problem it is trying to solve. For example, an AI algorithm for finding cancer-causing genes does not need facial recognition capabilities; however, when asked to find a dozen faces among a large crowd, it does not need to understand anything about gene interactions. The achievement of AGI simply means that a single algorithm performs multiple tasks, not that it does everything at once.

It is worth noting that AGI is different from ASI. ASI has certain human capabilities and perception, self-awareness, and can think independently and solve problems. Although both concepts seem to correspond to the problem-solving capabilities of AI, AGI is more like an omnipotent computer. At the same time, ASI transcends the technical attributes to become something like a human wearing an Iron Man suit. Oxford philosopher and leading AI thinker Nick Bostrom defines ASI as "an intelligence that is smarter than the best human brains in practically every field, including scientific creativity, general wisdom, and social skills."

2.2.3 Achieving AGI

Since the birth of AI, scientists have been working to achieve AGI, which can be divided into two specific paths.

The first path is to make computers surpass humans in some specific tasks, such as playing Go or detecting cancer cells in medical images. If computers can outperform humans in executing some difficult tasks, then it is eventually possible for computers to be stronger than humans in all tasks. Achieving AGI in this way, the working principle of AI systems and the flexibility of computers become irrelevant.

The only important thing is that such AI computers perform specific tasks better than other AI computers and ultimately surpass the strongest humans. If the strongest computer Go player is only ranked second worldwide, it will not make headlines and may even be considered a failure. However, defeating the world's top human players would be seen as significant progress.

The second path focuses on the flexibility of AI. In this way, AI does not have to perform better than humans. Scientists aim to create machines that can do various things and apply what they have learned from one task to another.

For example, the rapidly developing AIGC follows this path. Over the past year, the progress of AIGC technology mainly includes three aspects: image generation represented by DALL·E 2 and Stable Diffusion, NLP based on GPT-3.5's ChatGPT, and code generation like Copilot based on CodeX.

Taking ChatGPT as an example, this popular AI language model quickly became a global sensation after its release. Unlike the previous AI voice assistants' silly responses, ChatGPT is unexpectedly smart. It is used to create stories, write news, answer objective questions, chat, write code, and find code issues, among other things. Media even claims that ChatGPT will be the next disruptor in the tech industry.

In fact, AI like ChatGPT is entering human society as a disruptor. Based on a vast dataset, ChatGPT has better language understanding capabilities, which means it acts more like a general task assistant, capable of integrating with different industries, becoming a top expert in various fields, and generating many application scenarios. It can be said that ChatGPT has opened a door for AGI.

Moreover, ChatGPT introduces human supervisors specifically to "teach" AI how to answer human questions better, allowing AI to optimize data and parameters according to human values. On the Internet, as long as it involves text generation and conversation, ChatGPT can "process" it, achieving a result that closely resembles natural human language dialogue.

Take autonomous driving as an example, current self-driving systems are still ANI systems, and their interactions with humans are relatively mechanical. For instance, when a vehicle is in front of them, the system might need help deciding when to pass according to the rules. However, the iteration of AI, like ChatGPT, might bring machines closer to human thinking patterns, learning human driving behavior and leading autonomous driving into the "era 2.0."

In conclusion, developing AI technologies like ChatGPT represents significant progress in pursuing AGI. By improving flexibility, learning from different tasks, and integrating with various industries, AI systems have the potential to become more adaptable and versatile, bringing us closer to the realization of AGI. While we may have yet to achieve true AGI, these advancements suggest we are moving in the right direction.

2.3 ChatGPT Revolutionizing AGI Potential

While in the past, people had various abstract ideas about AGI, today, with the development of AI generation technologies such as image generation, code generation, and NLP, AGI seems to have reached a crucial crossroads. AIGC is a technological foundation for revolutionary scenarios, encompassing applications such as graphic and text creation, code generation, gaming, advertising, and artistic graphic design.

The popularity of ChatGPT and the opening of the ChatGPT API (Application Programming Interface) are pushing large-scale pretraining of multimodal models, AIGC, and other AI technologies represented by the edge of the upcoming explosion, and the human imagination of AGI is beginning to take shape.

2.3.1 ChatGPT's Open API

On March 1, OpenAI officially announced the opening of the ChatGPT API, which means that developers can now integrate ChatGPT and Whisper models into their applications and products through the API. This means that businesses or individual developers no longer need to develop ChatGPT-like models themselves but can directly use models such as ChatGPT for secondary applications and development.

APIs are actually application programming interfaces designed to facilitate smooth communication between two different applications, often referred to as the "middleman" of the application. In fact, in our daily lives, we often encounter hardware interfaces, the most common being HDMI and USB interfaces, which we know provide certain functions when connected. Like hardware interfaces, program interfaces can encapsulate the functions implemented within a program, making the program like a box with only one opening. People can use this function by accessing this opening. The person making the call easily use these functions without knowing the specific implementation process of these functions, and the API interface is used to call these functions according to the process specified by the author.

For example, when we go to a store to order food by scanning a QR code, we first need to scan the QR code to enter the page, enter the number of dinners, and then place an order. After placing the order, the waiter will check the menu with you and synchronize it with the kitchen, and then we wait for the food to be served. The ordering process can be seen as the working process of the API interface. We select the dishes through an ordering API interface, let the waiter know our needs, and provide the corresponding food and service in the background. This process is the role of the ordering API interface.

Before the OpenAI API was opened, people could communicate with ChatGPT, but they needed help to develop further applications based on ChatGPT. However, on March 1, OpenAI officially announced that developers can now integrate ChatGPT and Whisper models into their applications and products through the API. The Whisper API is an AI-driven speech-to-text model launched by OpenAI in September last year.

Specifically, the ChatGPT API is supported by the AI model behind ChatGPT, which is called GPT-3.5-turbo. According to OpenAI, it is faster, more accurate, and more powerful than ChatGPT and GPT-3.5. The pricing of ChatGPT API is $0.002 per 1,000 tokens (about 750 words), which is 90% cheaper than the existing GPT-3.5 model. The reason why ChatGPT API can be so cheap is partly due to "system-wide optimization."

OpenAI states this will be much cheaper than using existing language models directly. Greg Brockman, OpenAI's President and Chairman said, "APIs have always been in our plans, but we needed time to get these APIs to a certain quality level." Now that the time has come, ChatGPT API is officially available to the public.

Several companies have already integrated ChatGPT API to create chat interfaces. For example, Snap has launched My AI for Snapchat+ subscribers, an experimental feature based on ChatGPT API. This customizable chatbot can provide advice and even write a joke for a friend in seconds. Currently, Snapchat has reached 750 million MAU.

Shopify has created an "intelligent shopping assistant" for its app Shop, which has a user base of 100 million, using the ChatGPT API. When consumers search for products, the AI will make personalized recommendations based on their requirements. The Shop AI assistant will simplify shopping by scanning millions of products to help users quickly find what they want.

Quizlet is a learning platform used by over 60 million students. Over the past three years, Quizlet has partnered with OpenAI to use GPT-3 in multiple use cases, including vocabulary learning and practice tests. With the launch of ChatGPT API, Quizlet has also released Q-Chat, an "AI teacher" that asks adaptive questions based on relevant learning materials and attracts students with a fun chat experience.

Currently, OpenAI is continuously improving its ChatGPT model and hopes to provide enhanced features to developers.

2.3.2 *The Generality of ChatGPT*

It is evident that the ChatGPT API, which supports many different applications, is a powerful tool. Previously, some developers tried to integrate OpenAI's regular GPT API into their applications but could not achieve the same level of effectiveness as ChatGPT. By officially opening the ChatGPT API, OpenAI has opened up a new door for developers.

Developing a chatbot model like ChatGPT is completely out of reach for most businesses and developers. According to SemiAnalysis, ChatGPT's one-time training cost is estimated to be $840 million, and the cost of generating a message is around 1.3 cents, which is three to four times that of traditional search engines. This is the cost of cultivating ChatGPT for OpenAI. OpenAI has almost gone bankrupt due to insufficient funds. The success of ChatGPT also sets a high entry barrier for newcomers who must have a solid AI foundation and ample funding. However, at a relatively low cost, the official release of the ChatGPT API opens the door for developers to build their own chatbots.

More importantly, the release of the ChatGPT API provides a realistic path for AGI. By the standard of whether it can perform multiple tasks, ChatGPT has the characteristics of AGI. ChatGPT was trained to answer various types of questions and is used in multiple application scenarios to simultaneously perform multiple tasks, such as question and answer, conversation generation, text generation, and more. This means that it is trained for a specific task and has a general language processing ability. Therefore, we can also regard ChatGPT as a AGI model.

The release of the ChatGPT API allows everyone to use this AGI model. It can be said that the ChatGPT API builds a complete underlying application system for the development of AI. This is like the operating system of a computer. The operating system is the core part of a computer and plays a vital role in resource management, process management, file management, and more. Regarding resource management, the operating system is responsible for managing the computer's hardware resources, such as memory, processor, disk, etc. It allocates and manages these resources so that multiple programs can share resources and run efficiently. In process management, the operating system manages the programs running on the computer, controls their execution order and resource allocation, maintains communication between programs, and handles concurrency issues between programs. In file management, the operating system provides a standard set of file systems that make it easy for users to manage and store files.

Android and iOS are currently the two mainstream mobile operating systems, and the emergence of the ChatGPT API provides a technical foundation for AI applications. Although ChatGPT is a language model, conversing with humans is only the surface of ChatGPT's capabilities. The real role of ChatGPT is to use the open interface based on the ChatGPT open-source system platform to make some secondary applications.

AI may become a basic infrastructure like water and electricity in the future. In 1764, a British textile worker named Hargreaves invented a spinning jenny, which could spin eight threads at once, greatly improving productivity. The appearance of the spinning jenny triggered a chain reaction of technological innovation in machine invention and opened the prelude to the industrial revolution.

As machine production increased, Britain led the way into the industrial revolution in the mid-18th century. At that time, the energy source for steam engines was coal, which was based on coal that far exceeded human energy so that human productivity could be greatly improved. Due to the increase in efficiency, the price of coal became very cheap. Today's AI is almost like the coal of that time because future intelligence will undoubtedly become society's most basic infrastructure and mass commodity. Just like we cannot live without water, electricity, gas, and the Internet in real life, the technical foundation of AI is the infrastructure of the intelligent era.

2.4 More Powerful Versions of ChatGPT

On March 15 of this year, GPT-4 made headlines. If the birth of ChatGPT was a bombshell in the field of AI, then the release of GPT-4 is a new thunderbolt that ignites our imagination of AI.

2.4.1 From a Test Product to GPT-4

In fact, while most people were amazed at the powerful capabilities of ChatGPT, few knew that ChatGPT was actually just a hastily released test product by OpenAI.

According to US media reports, in mid-November 2022, OpenAI employees were asked to launch a chatbot quickly. An executive said the chatbot would be called "Chat with GPT-3.5" and open to the public for free in two weeks. This was different from the original plan. OpenAI has been developing a more powerful language model called "GPT-4" for the past two years and planned to release it in 2023. GPT-4 was being internally tested and fine-tuned in 2022 to prepare for its launch. However, OpenAI's executives changed their minds.

Due to concerns that competitors might release their own AI chatbots that could surpass them before GPT-4 was released, OpenAI brought out an enhanced version of the old language model GPT-3 called GPT-3.5 and made adjustments based on it. This led to the birth of the new chatbot ChatGPT.

Admittedly, although ChatGPT has given us a glimpse of the prototype of universal AI, it still needs to overcome many-objective problems. In some professional fields, ChatGPT still makes some low-level mistakes, especially in fields with formulas and calculations, such as mathematics, physics, and medicine.

Unlike the hasty release of ChatGPT, GPT-4 was well prepared. According to online rumors, GPT-4 was trained and completed in August last year. It is only now being released because OpenAI needed six months to make it more secure. GPT-4's three main features are image recognition, advanced reasoning, and a vast vocabulary.

Regarding image recognition, GPT-4 can analyze images and provide relevant information, such as recommending recipes based on photos of ingredients and generating image descriptions and captions. However, due to concerns about potential abuse, OpenAI has delayed the release of image description functionality. That is to say, the image input function of GPT-4 is still in the preview stage and not yet publicly available and can only be seen in OpenAI's live demonstrations.

As for advanced reasoning, GPT-4 can schedule a meeting time for three people based on different situations and answer complex questions with contextual relationships. For example, if you ask what happens when the rope in the picture is cut, it will answer that the balloon will fly away. GPT-4 can even tell some low-quality, patterned jokes. Although they are not very funny, it has begun to understand the human trait of "humor." It is worth noting that the reasoning ability of AI is a sign of its slow evolution toward human thinking when the complexity of the task reaches a sufficient threshold.

Regarding vocabulary, GPT-4 can handle 25,000 words, eight times that of ChatGPT, and write code in all popular programming languages. In casual conversation, the difference between ChatGPT and GPT-4 is subtle. However, when the complexity of the task reaches a sufficient threshold, the difference appears. GPT-4 is more reliable, more creative, and handles more subtle commands than ChatGPT.

Moreover, GPT-4 is also expected to significantly impact various industries and fields, such as education, healthcare, and finance. In the field of education, GPT-4 could be used as a tutor, providing personalized learning experiences and adapting to the student's level of understanding. In healthcare, GPT-4 could assist doctors in diagnosing diseases and recommending treatment options, improving patient outcomes. GPT-4 could be used in finance for predictive modeling and risk analysis, helping financial institutions make better investment decisions.

However, with the development of more powerful AI models such as GPT-4, there are also concerns about their potential negative impacts. Some experts warn that these models could be used maliciously, such as creating fake news, deep fake videos, or even more sophisticated cyberattacks. Others worry about the ethical implications of creating AI systems that are too advanced, with the potential to outsmart their human creators.

Despite these concerns, the development of GPT-4 represents a significant step forward in the field of AI, pushing the boundaries of what is possible and inspiring innovations and breakthroughs. As AI technology continues to evolve, it will be crucial to balance its potential benefits with its potential risks, ensuring that these powerful tools are used for the greater good of humanity.

2.4.2 The Differences between GPT-4 and ChatGPT

Besides having better performance than ChatGPT, what are the differences between GPT-4 and ChatGPT?

OpenAI claims they spent six months making GPT-4 safer than the previous version. The company improved the monitoring framework and worked with experts in sensitive areas such as medicine and geopolitics to ensure the accuracy and safety of GPT-4's answers. GPT-4 has more parameters, meaning it will be closer to human cognitive performance than the previous version.

According to the OpenAI website, the biggest evolution of GPT-4 compared to ChatGPT is "multimodality" and long-form content generation. The key here is "multimodality," which means diffusing different data types. Those who have used ChatGPT may have noticed that its input type is pure text, while the output is language text and code. GPT-4's multimodality means that users can input different types of information, such as video, audio, images, and text. Similarly, GPT-4, with multimodal capabilities, can generate videos, audio, images, and text based on the information provided by users. Even if both text and images are sent to GPT-4, it can generate text based on these two different types of information.

Another major focus of the GPT-4 model is the establishment of a predictably scalable deep learning stack. Because extensive specific model adjustments are not feasible for large training like GPT-4, the OpenAI team developed infrastructure and optimization with predictable behavior at multiple scales. To validate this scalability, researchers accurately predicted the final loss of GPT-4 on the internal codebase in advance, using models trained with the same method but with a computing volume of 1/10000.

Although GPT-4 is more powerful, it still has similar limitations to early GPT models: it is still unreliable and has factual "illusions" and reasoning errors. It is important to be careful when using language model outputs, especially in high-risk contexts, and use specific protocols that fit the specific use case. However, GPT-4 significantly reduces illusions compared to previous models. The likelihood of responding to unauthorized content requests is reduced by 82%. In OpenAI's internal adversarial authenticity evaluation, GPT-4 scored 40% higher than GPT-3.5.

After OpenAI released GPT-4, its partner and investor Microsoft immediately responded, saying, "The new Bing is running on GPT-4, which we've customized for search." Obviously, as OpenAI updates GPT-4 and higher versions, Bing also benefits from these improvements.

In addition, OpenAI announced partnerships with language learning application Duolingo and the company behind the app Be My Eyes, designed specifically for the visually impaired to support people with disabilities. The US nonprofit educational institution Khan Academy will use GPT-4 to create AI tutors for students; the Icelandic government will use it to help maintain the Icelandic language; and financial company Morgan Stanley will use GPT-4 to manage, search, and organize its massive content library.

Looking further into specific applications, GPT-4's advanced reasoning skills can provide more accurate and detailed answers for users. Given GPT-4's stronger language and image recognition capabilities, it can simplify the creation of marketing, news, and social media content. In education, GPT-4 can help students and educators by generating content and answering questions in a more human-like way.

2.4.3 The Release of GPT-4

In fact, if we only consider the ability to perform multiple tasks as a standard, the previously released ChatGPT already had the characteristics of a AGI. ChatGPT was trained to answer various types of questions and could be applied to multiple scenarios, including question answering, dialogue generation, and text generation. This shows that ChatGPT has been trained for a specific task and has general language processing capabilities.

More importantly, the success of ChatGPT proves the effectiveness of the large model approach. Before OpenAI's GPT models, people used RNNs for NLP and added attention mechanisms. However, RNN + Attention prolongs the entire model's processing speed because RNN processes one word at a time.

Therefore, the Google team later proposed "No RNN, Only Attention." This natural language model with no RNN and only attention is the Transformer model, which is the technical basis for the success of ChatGPT today. The Transformer model with only attention no longer processes one word at a time but processes one sequence at a time, which is computed in parallel, greatly increasing the speed of computation and making it possible to train large, super large, huge, and super huge models.

With time and cooperation with professional organizations in specific fields, using ChatGPT's learning capabilities and optimized parameters and models, GPT-4 will soon become an expert-level model in some professional fields. GPT-4 is essentially a more powerful version of ChatGPT that has been further trained and optimized.

Just like our human thinking and learning, we can generate novel ideas and insights by reading a book. Humans have absorbed vast amounts of data from the world, which has changed the neural connections in our brains in countless ways. LLMs in AI research can do similar things and effectively guides their intelligence.

With the release of ChatGPT to the public and millions of people interacting with it, ChatGPT will gain a huge amount of valuable data. With its more powerful learning ability than humans, its learning and evolution speed surpasses our imagination.

With the release of more powerful GPT-4 and even future generations, and OpenAI making its technology a general underlying AI technology for use in all industries, AI can quickly acquire professional knowledge in various fields. With various international scientific journals and research materials, AI can also provide analysis, suggestions, models, and simulations for scientists' research, even conducting simulated scientific research.

2.4.4 *The Victory of the Large Model*

In addition to being able to perform multiple tasks and secondary applications, more importantly, the success of ChatGPT and GPT-4 has proven the effectiveness of the large model. This has directly opened the door to the development of AGI, allowing AI to finally achieve the breakthrough from 0 to 1 and usher in the true AI era.

The success of ChatGPT and GPT-4 not only lies in their ability to perform multiple tasks and applications but also in the effectiveness of the large model approach. Before OpenAI's GPT model, NLP models were built using RNNs and attention mechanisms. However, the RNN + Attention approach was slow since RNN processes word one at a time. In 2017, the Google Brain team proposed the Transformer model, which only uses attention mechanisms, making it much faster and able to process sequences in parallel.

OpenAI developed the first-generation GPT model, which had 117 million parameters and could predict the next word in a sentence. With the use of the Transformer model and large datasets, GPT was able to generate coherent paragraphs of text. At the same time, Google's BERT model, also based on the Transformer architecture, achieved excellent results in sentence-level semantic analysis through a special training method of pretraining and fine-tuning.

Despite the popularity of BERT, the developers of GPT continued to focus on generative models rather than understanding, leading to the development of GPT-3, which was able to complete various tasks involving text output. From GPT-1 to GPT-3, OpenAI spent more than two years increasing the number of parameters from 117 million to 175 billion, proving the feasibility of large models and the relationship between the number of parameters and AI capabilities. However, after the success of GPT-3, instead of continuing to increase the number of parameters, the developers spent nearly two years and many resources incorporating human feedback and reinforcement learning into the GPT series, optimizing the data and parameters according to human values.

The success of ChatGPT and GPT-4 is not only a technological achievement but also an engineering triumph that validates the large model approach. It has paved the way for the development of genuinely versatile AI, transitioning from a nascent stage to full-fledged capability, ushering in a new era of AI.

2.4.5 Challenges of Large Models

Although the rapid development of AI based on the large model technology path has given us hope for the development of AGI, it still needs to have a fundamental breakthrough. The reason behind the success of today's AI generation is the large training datasets. This means that without fundamental innovations, AGI may continue to be developed through larger and larger models. ChatGPT is an example of how combining a large dataset with a powerful Transformer model leads to a deep understanding of natural language. Although the exact size of ChatGPT's neural network has not been disclosed, its predecessor, GPT-3, already had 175 billion parameters.

However, the increasing size of models also brings some challenges. On the one hand, there may need to be more available computing resources to support the maximization of model size. As data explodes and computing power develops rapidly, a high-energy world emerges. With the increase in computing power, there is also a corresponding increase in the electricity demand. For example, GPT-3 consumes enormous computing power, requires about 190,000 kWh of electricity, and generates 850,000 tons of carbon dioxide per training cycle, making it an "energy monster." According to incomplete statistics, in 2020, about 5% of global electricity generation was consumed by computing power, which could increase to 15% to 25% by 2030. This means that the power consumption of the computing industry will be comparable to that of heavy energy-consuming industries. In fact, for the computing industry, electricity costs are the most important costs after chip costs.

On the other hand, large models may not be able to scale up for some important tasks because, with cognitive models and common sense, large models are easier to reason with. When Google released its version of ChatGPT called Bard, it made a factual error in its first online demo video. In the video, Bard answered a question about James Webb Space Telescope's discoveries and claimed it had "taken the first pictures of planets outside the solar system." This is incorrect. The first photo of a planet outside the solar system, also known as an exoplanet, was taken by the Very Large Array (VLA) in Chile in 2004.

An astronomer pointed out that the problem may be due to AI misunderstanding a "vague NASA news release." Bard's error highlights a larger problem with AI-driven search: AI can confidently make factual errors and spread incorrect information—they do not "understand" the information they are conveying but instead make guesses based on probabilities. In fact, Google and Microsoft acknowledge that ChatGPT-based chat services face similar challenges. If the model only learns syntax and semantics but fails in pragmatics or common sense reasoning, then we may not be able to obtain trustworthy AGI.

2.5 The Foreseeable Future of ChatGPT

In mathematics, "singularity" describes situations where normal rules no longer apply, similar to asymptotes. In physics, singularity describes phenomena such as an infinitely small, dense black hole or the critical point we were squeezed into before the big bang, where the usual rules no longer apply.

In 1993, Vernor Vinge wrote a famous article that used this term to describe the moment when our intelligent technology surpassed our capabilities. For him, at that moment, all of our lives will be forever changed, and the usual rules will no longer apply.

Now, with the explosion of ChatGPT, we seem to be on the eve of a technological singularity.

2.5.1 ChatGPT's Path to Surpassing Human Capabilities

In fact, the greatest feature of AI is not just a change in the field of the Internet nor a disruptive technology for any particular industry. Still, as a universal technology that supports the entire industry structure and ecological economic changes, it is one of the important tools that can project its energy into almost all industry sectors, promote their industrial transformation, and provide new impetus for global economic growth and development. From ancient times to now, no technology like AI has sparked unlimited human imagination.

Since AI is not a single technology, its coverage is pervasive, and the meaning represented by the word "intelligence" can almost replace all human activities. Even if it is only a "human-level" intelligent technology, what AI can do greatly exceeds people's imagination.

In fact, AI has already covered all aspects of our lives, from spam filters to ride-hailing apps. Algorithms generated by AI recommend the news we read every day. Online shopping websites display the products most likely to interest and be purchased by users based on AI recommendations. This includes increasingly simplified automated driving vehicles, facial recognition-based clock-in systems for work, and so on, some of which we deeply feel, while others quietly infiltrate society's mundane daily operations. Everything we experience now is still in the ANI stage, where all AI products in our lives can only perform single tasks.

But the arrival and explosion of ChatGPT have pushed AI onto a high-speed track of application. Although AI has greatly changed the current era, ANI products still have many limitations and "unintelligent" areas. For example, whether it is Alibaba, Baidu, or JD.com (JD) if you encounter their AI customer service, you will basically be turned into an idiot and unable to have a pleasant chat, let alone solve problems normally. However, ChatGPT has a humanoid logical ability, and the powerful chat we currently see is only this stage based on data updates from 2021.

Moreover, repetitive language and text work does not really require complex logical thinking or top-level decision-making judgment. For example, answering the phone or handling email, helping customers book hotels and meals through language and text work, filling in data and information into contracts, financial reports, market analysis reports, factual news reports that are presented according to fixed formats, extracting outlines from existing text materials, organizing key points, converting real-time text records of meetings into briefings, writing process-oriented, procedural articles, and so on. These works are all scenarios where products based on ChatGPT or other large models can be applied.

Overall, currently, ChatGPT has already shown its creativity, whether it is AI dialogue, AI writing, or AI painting. The inherent non-determinism, divergence, and free-spiritedness of large-scale pretrained models are good helpers to inspire human inspiration. In the future, with the updates and iterations of ChatGPT, market work that requires creating advertising copy or business presentations, movie screenwriting work that requires exploring different storylines divergently, game scene design work that requires greatly enriching visual experiences, etc., may also be full of the shadow of ChatGPT.

Lee Kai-Fu once mentioned a point of view that AI would replace work that is at most 5 seconds of thinking in the future. Looking at it now, in some fields, ChatGPT has already far exceeded this standard of "five-second thinking." With its continued evolution, coupled with its powerful ML ability and quick learning and evolution in our human interaction process, replacing and surpassing us humans in all regular and rule-based work fields is only a matter of time.

2.5.2 *The Eve of the Technological Singularity*

Human progress is accelerating over time—this is the Law of Accelerating Returns described by futurist Ray Kurzweil. This happens because more advanced societies can progress faster than less developed ones, as they are more advanced. People in the 19th century knew more and had better technology than those in the 15th century, so the progress made in the 19th century was much greater than in the 15th century.

For example, in the 1985 film *Back to the Future*, the "past" occurred in 1955. In the movie, when Michael J. Fox returns to 1955, he is taken aback by the novelty of television, the price of soda, the screeching electric guitar, and the slang changes. It was a different world. But it might be even more interesting if the movie was made today and the "past" was set in 1993. Any of us traveling back to a time before the widespread adoption of mobile Internet or AI would be even more out of place and ill-adapted than Michael J. Fox was in 1955. This is because the average rate of progress from 1993 to 2023 was higher than the rate from 1955 to 1985—because the former is a more advanced world—there have been many more changes in the past 30 years than in the previous 30 years.

Futurist Kurzweil believes: "In the first tens of thousands of years, the pace of technological growth was so slow that one generation could not see significant results; in the past century, a person could see at least one major technological advancement in their lifetime; and since the 21st century, changes similar to the sum of all previous human technological advancements will occur roughly every three to five years." In short, due to the Law of Accelerating Returns, Kurzweil believes that the 21st century will see 1,000 times the progress of the 20th century.

Indeed, the pace of technological progress has even exceeded the limits of individual comprehension, and ChatGPT—born in an era of rapid technological change—has immense potential.

In September 2016, after AlphaGo defeated the European Go champion, many industry scholars and experts, including Lee Kai-Fu, believed that AlphaGo had little hope of further defeating world champion Lee Sedol. However, the result was that just six months later, AlphaGo easily defeated Lee Sedol and remained undefeated after losing one game, a pace of evolution that left people speechless.

Now, AlphaGo's evolution speed may be replayed with ChatGPT. ChatGPT is built on OpenAI's GPT-3.5 model. Since 2018, GPT-1, GPT-2, and GPT-3 have had parameters of 117 million, 1.5 billion, and 175 billion, respectively. This is exponential growth, and the latest GPT-4 performance will be even more powerful, reaching a height that is difficult for us to imagine today.

Although ChatGPT does have limitations at this stage. It is an imperfect AI product with bugs, this still cannot deny the significance of ChatGPT—the AI that human society has been discussing for years has finally evolved from artificial idiocy toward the envisioned AI.

The singularity is looming, and the future has arrived. As Kevin Kelly, one of the most famous Internet prophets and known as the "Father of Silicon Valley Spirit," said: "It took only 58 years from the first chatbot (ELIZA, 1964) to a truly effective chatbot (ChatGPT, 2022). So, don't assume that it must be clear because something is close in time, and don't assume that it must be impossible just because something is far away."

3

Commercial Battles of ChatGPT

3.1 ChatGPT's Rise: Transforming OpenAI's Finances and Valuation

With the sudden rise of ChatGPT, its parent company OpenAI has also received widespread attention worldwide. In fact, before the advent of ChatGPT, OpenAI was a company that was losing money. In 2022, the company had a net loss of $540 million, and as the number of users increased, its computing costs increased, which could further expand the losses. OpenAI co-founder and CEO Sam Altman responded to Musk's question about cost on Twitter in December, saying that each ChatGPT conversation costs a few cents.

However, the explosion of ChatGPT suddenly broke OpenAI's loss situation and showed great commercial potential. OpenAI's valuation also soared to $29 billion, doubling its valuation of $14 billion in 2021 and nearly 300 times higher than its valuation seven years ago.

3.1.1 The Legendary Life of ChatGPT's Father

The success of ChatGPT cannot be separated from OpenAI's CEO, Sam Altman. As an important figure in the rapid growth of ChatGPT's MAU to 100 million and OpenAI's valuation, Altman has also received more market attention. Many media have described Altman as the "outstanding person of the year" and the "father of ChatGPT."

Sam Altman's experience can be described as legendary. Born in Chicago, Illinois, on April 22, 1985, Altman grew up in St. Louis, Missouri. Altman showed talent in computers at a very young age and mastered the system behind the code in his kindergarten district when he was still a child.

At the age of 8, Altman got his first personal computer and developed a strong interest in programming. He even disassembled an Apple Macintosh computer, which became an important connection between him and the world. For example, he discovered that AOL's online chat rooms were a disruptive innovation for information retrieval and socializing.

After graduating from high school, Altman entered Stanford University to study computer science. He was unwilling to concentrate on studying and was determined to start a business. In his sophomore year, Altman and his classmates founded Loopt, a mobile application that allows friends to share their location information. In 2005, 19-year-old Altman and his classmates became the first group of entrepreneurial teams to enter YC. Later, Altman dropped out of school with two classmates and devoted himself to Loopt. At that time, location-based services were top-rated, and Altman was fortunate to receive investment from Sequoia Capital. Loopt raised five rounds of funding in four years, totaling $39.1 million. However, Loopt needed to attract more consumers.

In October 2009, 25-year-old Altman sold the company to Graffiti Geo for $43 million, which caused the venture capital firms that invested in them to lose some money, but it was the best arrangement for them. Because three years later, in 2012, the product was closed due to an inability to continue operations. Altman himself received a $5 million return.

After selling the company, Altman did not start the next business but took a rest for more than a year. However, this year and a half had a profound and important impact on Altman's future. During that year, Altman learned much knowledge in the fields of interest, such as nuclear engineering, AI, and synthetic biology, and began understanding the situation of the four projects he had previously invested in as an angel investor.

In 2011, Altman began working part-time at YC. He founded a small venture capital fund called Hydrazine Capital, which raised $21 million, of which he invested $5 million himself, including a large investment from Peter Thiel, co-founder of PayPal. 75% of the fund was invested in YC companies. As it turned out, Altman was very good at investing. For example, Altman led the B-round financing of Reddit. This company had long been chaotic and disordered after graduating from YC, and he served as CEO for eight days before inviting the founder back to continue as CEO. Due to YC's high success rate in incubating projects, Altman's strategy was a great success. In just four years, the value of Hydrazine Capital increased tenfold.

In 2014, at the age of 28, Altman took over as President of YC, becoming a well-known figure in Silicon Valley. Under Altman's leadership, YC continued to move forward in the direction he expected.

Altman is also a board member or adviser for several companies, including OpenDoor, Postmates, and RapidAPI. He has helped these companies secure tens of millions of dollars in investments and played a crucial role in their successful public listings. Altman is a Senior Fellow at Carnegie Mellon University and has published numerous articles on technology innovation and entrepreneurship. He has extensive experience in entrepreneurship, investment, and technology and is highly praised for his exceptional talent.

In 2015, at the age of 29, Altman was selected for *Forbes*'s list of the Top 30 Venture Capitalists Under 30. It was also this year that Altman co-founded the non-profit organization OpenAI with Tesla CEO Elon Musk.

3.1.2 Non-profit Organization OpenAI

It's hard to believe that today's global unicorn OpenAI started as a non-profit organization. The story of OpenAI's origin is quite dramatic.

In 2014, Google acquired DeepMind for $600 million. Considering that Google's DeepMind was the first company most likely to develop AGI, Elon Musk said that if humans developed AI with some bias, there would be an eternal, superpowerful dictator. Even a small personality flaw could make its first step to kill all AI researchers. In other words, if DeepMind succeeds, it might use extreme means to monopolize this all-powerful technology. Therefore, Musk and others believed that a laboratory competing with Google was needed to ensure this did not happen. And this competing laboratory with Google was later the non-profit organization OpenAI.

In December 2015, OpenAI was established in San Francisco, raising $1 billion in funds, with main sponsors including Tesla's founder Musk, PayPal's co-founder Peter Thiel, LinkedIn's founder Reid Hoffman, YC's President Sam Altman, Stripe's CTO Greg Brockman, Y Combinator co-founder Jessica Livingston; and some institutions such as YC Research, Altman's foundation, Infosys, and Amazon Web Services (AWS).

OpenAI's establishment aims to achieve AGI and build a mechanical system that learns and reasons like human minds. Since its establishment, OpenAI has also been engaged in basic AI research. In fact, before the birth of ChatGPT, many people may not have heard of this company.

However, soon the founders of OpenAI found that having an ideal to benefit humanity was far from enough—maintaining non-profit nature could not sustain the normal operation of the organization because once the research is conducted, breakthroughs require computing resources that need to double every 3–4 months, which requires funding to match this exponential growth. And the non-profit nature of OpenAI at that time was also very restrictive and far from achieving self-reliance. The money burning problem was also validated with DeepMind. After being acquired by Google that year, DeepMind did not generate profits for Google in the short term but instead burned several hundred million dollars from Google each year. In 2016, it lost 127 million pounds. In 2017, it lost 280 million pounds; in 2018, the loss reached as high as 470 million pounds, and the rate of burning money increased yearly. Perhaps due to this ideological conflict, Musk resigned from OpenAI's board of directors in February 2018, claiming to avoid conflicts with Tesla's operations and continue donating and advising this non-profit organization.

To solve the funding problem, in March 2019, Sam Altman stepped down as YC president and became chairman while also becoming CEO of OpenAI, concentrating more on OpenAI. Under Altman's leadership, OpenAI established a restricted profit entity—OpenAI LP, a hybrid of profit and non-profit called the "profit cap" by OpenAI.

According to OpenAI's statement in March 2019, if OpenAI could complete its mission of ensuring that AGI benefits all humanity, investors, and employees could receive returns limited by the cap. Under this new investment framework, the first round of investor returns were designed not to exceed 100 times, with subsequent rounds of returns lower. This unusual structure limits investor returns to multiples of their initial investment.

From this time point, the term "OpenAI" officially referred to the for-profit entity of OpenAI, namely OpenAI LP, instead of the original non-profit entity, OpenAI Inc. OpenAI LP is supervised by the board of OpenAI Inc to address the organization's needs for computing power, funding, and talent. Any excess returns will be donated to OpenAI Inc, the non-profit entity of OpenAI.

3.1.3 The Cooperation with Microsoft

In July 2019, the newly restructured OpenAI company received a $1 billion investment from Microsoft. Since then, OpenAI has been working closely with Microsoft, fulfilling its promised investment in 2019 and investing again in 2021. However, the funding is just the first layer of the cooperation between the two companies, which has proven to be a win-win partnership.

On the one hand, OpenAI urgently needed computing power investment and commercial endorsements. Sam Altman made several efforts to persuade Microsoft to invest, flying to Seattle to meet with Microsoft CEO Satya Nadella after taking over OpenAI LP. On the other hand, as Google's direct competitor, while Google continued to invest in AI, Microsoft's AI technology had been dwindling in commercial applications, especially after the failure of the Tay chatbot in 2016. Thus, Microsoft needed technological breakthroughs to regain its AI competitiveness.

According to Altman, after Microsoft's initial investment in OpenAI in 2019, the funds would be used to accelerate the development and commercialization of AGI. OpenAI would also use Microsoft's Azure as its exclusive cloud computing provider, and both companies would jointly develop new technologies and features. Reportedly, OpenAI spends about $70 million annually on model training on Microsoft's cloud service, which constitutes a significant part of Microsoft's investment in OpenAI. This is a win-win partnership, with Microsoft becoming OpenAI's "preferred partner" for technology commercialization and gaining exclusive authorization to use OpenAI's technological achievements. OpenAI, on the other hand, could leverage Microsoft's Azure cloud service platform to solve commercialization issues and alleviate high-cost pressures.

With the backing of Microsoft's cloud, OpenAI's computing power and confidence have grown, leading to its breakthrough achievement, GPT-3, in 2020. The same year, Microsoft acquired exclusive licensing for GPT-3's underlying technology. It gained priority authorization for technology integration, using GPT-3 in products such as Office, Bing, and Microsoft Design to optimize existing tools and improve product functionality.

In 2021, Microsoft invested in OpenAI again, and as the exclusive cloud provider for OpenAI, it deployed various tools developed by OpenAI, including GPT, DALL·E, and Codex, in Azure. This formed OpenAI's earliest source of revenue—providing paid APIs and AI tools to enterprises through Azure. Meanwhile, with the new commercialization authorization, Microsoft integrated OpenAI's tools with its products deeply and launched corresponding products. For example, in June 2021, in collaboration with OpenAI and GitHub, Microsoft launched the AI code completion tool GitHub Copilot based on Codex. The product was officially launched in June of the following year, offering services at a monthly fee of $10 or an annual fee of $100.

Entering 2023, with the explosion of ChatGPT, OpenAI and Microsoft announced an expansion of their partnership. According to the *Information*, Microsoft will invest up to $10 billion in OpenAI, and after the initial investors in OpenAI have recovered their initial capital, Microsoft will have the right to receive 75% of OpenAI's profits until it recovers its $13 billion investment, including the previous $2 billion investment in OpenAI as disclosed by *Fortune* magazine in January of this year. After this software giant earns $92 billion in profits, Microsoft's share will drop to 49%. At the same time, other investors and OpenAI employees will have the right to receive 49% of OpenAI's profits until they earn about $150 billion. Microsoft and investors' shares will be returned to OpenAI's nonprofit foundation if these limits are reached. Essentially, OpenAI is borrowing the company to Microsoft, and the length of the borrowing depends on how quickly OpenAI makes money. This means that Microsoft and OpenAI are further deepening their ties.

According to *Fortune* magazine, in 2022, OpenAI's revenue is estimated to be less than $30 million, while the total net loss is as high as $545 million, not including employee stock options. The release of ChatGPT could also quickly increase OpenAI's losses. In December, Altman responded to Musk's question about costs on Twitter, saying that each conversation with ChatGPT costs a few cents. After this round of funding, OpenAI, which was founded in 2015, is now valued at $29 billion, doubling its valuation of $14 billion in 2021 and increasing nearly 300 times from its valuation seven years ago.

3.1.4 *The Growth of OpenAI*

Although OpenAI has been burning money and making losses, there is no denying that since its establishment, it has achieved unprecedented breakthroughs in the field of AI.

On June 11, 2018, OpenAI announced an algorithm that performed well on many NLP tasks, which was the first-generation of GPT. GPT was the first algorithm to combine transformers with unsupervised pretraining, and its performance was better than the current known algorithms. This algorithm was the precursor to OpenAI's LLM exploration and paved the way for the more powerful GPT series.

Also, in June 2018, OpenAI announced that their OpenAI Five had begun to defeat amateur human teams in the game Dota2 and stated that they would compete with the world's top players in the next two months. OpenAI Five used 256 P100 graphics processing units (GPUs) and 128,000 central processing unit (CPU) cores to play 180 years of games every day to train its model. In the following months, more details about OpenAI Five were revealed. In professional matches in August, OpenAI Five lost two games to top players, but the model performed well in the first 25–30 minutes of the game. OpenAI Five continued to develop and announced on April 15, 2019, that it had defeated the then-current Dota2 world champion.

On February 14, 2019, OpenAI announced the GPT-2 model. The GPT-2 model had 1.5 billion parameters and was trained on 8 million web pages. GPT-2 was the scaled-up version of GPT, trained on data that was more than 10 times larger with more than 10 times more parameters.

On April 25, OpenAI announced its latest research result: MuseNet. This deep neural network can generate four-minute music pieces using ten different instruments and combining styles from country to Mozart to the Beatles. This marked OpenAI's expansion of generative models from the NLP field to other fields.

On May 28, 2020, OpenAI's researchers officially released the research results related to GPT-3, which was the world's largest pretrained model at the time with 175 billion parameters. Compared to GPT-2, released in February 2019, GPT-3 had significantly improved model capabilities, ease of use, and security, with better performance in tasks such as writing and summarizing text, translation, and dialogue.

On June 17 of the same year, OpenAI released the Image GPT model, which introduced the success of GPT to the field of computer vision (CV). The researchers believed that transformers were domain-agnostic, and they modeled sequences, so the field of CV could still use them. Image GPT also achieved good results at the time.

On January 5, 2021, OpenAI released CLIP, which can effectively learn visual concepts from natural language supervision. CLIP can be applied to any visual classification benchmark by providing the names of visual categories to be identified, similar to the "zero-shot" capability of GPT-2 and GPT-3.

On the same day, OpenAI released the DALL·E model, which was a hugely influential model. DALL·E was a 12 billion parameter version of GPT-3 trained on a dataset of text-image pairs to generate images from textual descriptions. DALL·E could create anthropomorphic versions of animals and objects, reasonably combine unrelated concepts, render text, and transform existing images. The release of DALL·E once again amazed the world.

On August 10, 2021, OpenAI released Codex, the successor to GPT-3. Codex was trained on both natural language and billions of publicly available source code lines, including code from GitHub repositories. Codex powers GitHub Copilot and is available as a paid API.

On January 27, 2022, OpenAI released InstructGPT, a language model that is even better at following user intent, making it more realistic.

On April 6, 2022, DALL·E 2 was released, with even more realistic and detailed results than the first version. After DALL·E 2 was officially opened for registration, the number of users exceeded 1.5 million, which doubled within a month.

On June 23, 2022, OpenAI trained a neural network to play Minecraft using a large unlabeled dataset of human gameplay videos with only a small amount of labeled data. The model can learn to make diamond tools. This task typically takes skilled humans over 20 minutes. It uses a human-native keyboard and mouse interface, making it highly versatile and representing a step toward agents for general-purpose computing.

On September 21, 2022, OpenAI released Whisper, a pretrained speech recognition model that approaches human-level performance and supports multiple languages. Unlike many other models with long-withheld results, Whisper is completely open-source, although its parameters are only 1.55 billion.

On November 30, 2022, OpenAI released the ChatGPT system, which performed nearly perfectly on many tasks and gained 1 million users in just 5 days. It helps with coding, technical explanations, multi-turn conversations, script writing, and legal documents. The birth of ChatGPT completely ignited the AI race and showed the powerful strength of OpenAI, its parent company.

3.1.5 The Confidence behind High Valuations

Currently, ChatGPT has become the undisputed unicorn company globally. Microsoft's plan to invest $10 billion in OpenAI has directly pushed OpenAI's valuation to $29 billion. The richness of OpenAI's revenue streams and the forward-looking nature of its investment portfolio indeed give OpenAI the confidence to have a high valuation.

Among them, subscription fees, API licensing fees, and commercial income generated by deep collaboration with Microsoft are currently the main sources of OpenAI's revenue.

Regarding subscription fees, on February 1, OpenAI announced the launch of a new pilot subscription plan called ChatGPT Plus, priced at $20 per month. The paid version offers benefits such as general access during peak periods, faster response times, and priority access to new features and improvements. OpenAI will continue to provide free access to ChatGPT alongside this subscription service. In the announcement, OpenAI mentioned that ChatGPT Plus is designed to ensure availability, faster response times, and priority access to new features even during high demand periods. Initially available to customers in the US, the company plans to gradually introduce paid subscription plans to users worldwide.

Even just using subscription fees brings significant revenue to OpenAI. Keep in mind that ChatGPT reached an astonishing number of 100 million MAU in just two months. If we look at the minimum charging standard, assuming 10% of people are willing to pay for it later, this has already brought OpenAI a potential annual revenue of $2.4 billion. And OpenAI's other application, DALL·E, which converts text into images, already had 1.5 million MAU in September 2022, plus its more professional use cases, leaving great room for the imagination.

API licensing fees charge other commercial companies to use models such as GPT-3. By integrating multiple large natural language models, mainly GPT-3, to gain a startup advantage, the most successful case is the AI writing unicorn company Jasper AI. The company's products are widely recognized in the industry, and Google, Airbnb, Autodesk, IBM, and others are its clients, with $75 million in annual revenue in 2022. A $125 million Series A financing was completed in October 2022, with a valuation of $1.5 billion, only 18 months after its product went online.

Another cooperation between GitHub, the world's largest open-source code hosting website, and OpenAI is the AI-assisted programming tool Copilot based on GPT-3. After starting to charge in June 2022, it had 400,000 subscribers in the first month, with a user payment rate of 1/3, far higher than that of general productivity software. It can be seen that even just the revenue channel based on API licensing fees still has a considerable potential market space to be explored.

In addition, the deep collaboration with Microsoft is another key point of OpenAI's commercialization. OpenAI and Microsoft believe that the former non-profit lab now has products that can be sold for-profit, and a commercialization path is essential.

On February 1, Microsoft launched advanced services embedded with ChatGPT in its collaborative work software Teams, which can automatically generate meeting notes, recommend tasks and personalized key content, and automatically divide meeting videos into multiple units based on topics, even if users miss the meeting, they can still obtain important personalized information. The price of Teams' advanced services is $7 per month.

On February 8, Microsoft announced the official launch of its new search engine Bing and Edge browser, supported by ChatGPT and GPT-3.5. Microsoft's market value surged over $80 billion overnight, reaching a new high in five months. The integration of ChatGPT into the Bing search engine is undoubtedly a significant move. For 13 years, Microsoft has been trying to compete with Google in the search engine market, but Bing's global market share has remained low, with Google holding over 90% of the market share while Bing only has a meager 3%. However, the release and popularity of ChatGPT in early December completely changed the situation. The two latest commercialization strategies will accelerate the business layout of GPT. One is that Microsoft is considering introducing ads in Bing, and the other is that OpenAI is starting to build a closed ecosystem.

In the face of ChatGPT-3 becoming a phenomenal product of AI, OpenAI's CEO boldly said that our future AI models would surpass human intelligence. Industry insiders predict that the scale of GPT-4 will reach 1 trillion parameters. Each person has 100 billion neurons and one million billion synapses. This means that the parameter quantity of the next generation of AI models will be equivalent to the synapses in the human brain. Once GPT-4 enters a higher variable mutation, its AI will become even more intelligent.

Currently, OpenAI is still a startup company in the red. However, OpenAI has expanded its industry-leading GPT auto-regressive language model to the commercial sector, becoming the fastest company to reach 1 million users in history. OpenAI predicts its revenue will grow rapidly as ChatGPT becomes an important tool for attracting customers. According to a document cited by the media, the company predicts revenue of $200 million in 2023 and a revenue of over $1 billion in 2024. OpenAI did not forecast its expenditure growth or when it could turn its losses into profits, but its potential for commercialization has put pressure on Internet technology giants.

3.2 Microsoft's Response

Microsoft is one of the tech giants closest to ChatGPT and its parent company, Open-AI. Thanks to its deep integration with ChatGPT, Microsoft seems to be the biggest winner at the moment, enjoying unparalleled prominence.

3.2.1 The Battle of Search Engines

For Microsoft, the greatest benefit is likely in the search business.

In the early hours of February 8, Beijing time, Microsoft officially launched the latest version of the Bing search engine and Edge browser, supported by ChatGPT. The new Bing search will answer questions with a similar context to ChatGPT. Microsoft CEO Satya Nadella said at the event, "AI will fundamentally change all software, starting with search, the largest category," and called it "a new day for the search" and "the race begins today."

Specifically, according to Microsoft's official website, users need to log in to their Microsoft account more quickly, to access the new Bing more quickly, set Bing as the default search engine, and download the Bing Search (English version) mobile app. The new Bing, integrated with ChatGPT, has two search modes. In one mode, traditional search results are displayed side by side with AI annotations. With the new Bing, users input queries of up to 1,000 words and receive AI-generated answers with annotations, which will appear alongside regular search results from the web. The other mode allows users to interact directly with ChatGPT, asking questions in a chat interface similar to ChatGPT, further optimizing answers, narrowing the scope, and providing more tailored answers to user needs.

According to Microsoft, the company expects to roll out access to millions of people in the coming weeks and launch a mobile version of the experience. According to the FAQ, users will receive an email when they are off the wait list and can access the new chat experience, successfully experiencing the new version of Bing.

Sam Altman, CEO of OpenAI, the parent company behind ChatGPT, also confirmed that the upcoming Microsoft product uses an upgraded AI language model, "Prometheus," which is more powerful than the GPT-3.5 currently used by ChatGPT. This means that the new Bing chatbot can provide consumers with brief introductions to current events, going beyond ChatGPT, which is currently limited to answering questions based on data up to 2021.

This is a completely new product form. It means that search engines are not just search engines but personalized search engines. Personalization, for example, when we want to create a diet plan focused on fat loss and muscle gain, we input our preferences in the chatbox, such as not liking celery, not wanting nuts, keeping calories within 800 kcal, and so on. After inputting these requirements, we can receive a personalized meal plan that meets our needs. This is just one daily example of personalization, but traditional search engines need help to achieve such customization features. Now, the Bing search engine with built-in ChatGPT is not only about helping users obtain information but also about helping them obtain information more efficiently and accurately.

In addition to upgrading the product form of the search engine, Bing with built-in ChatGPT has also impacted the search engine market. Over the past decade, Google has maintained absolute dominance in the international search market. According to Statcounter data, in January this year, Google's global search engine market share was as high as 92.9%, while Bing had only 3.03%. In the US market, Google's share was as high as 88.11%, with Bing at only 6.67%. Over the past decade, Google's market share in the US has grown from 81% to 88%, while the formerly second and third-ranked Bing and Yahoo have increasingly shrunk. The decline of the latter two seemed unstoppable before the ChatGPT craze.

Moreover, in the 12 months ending June, Microsoft earned $11.6 billion in advertising revenue from search, MSN, and other news products, up 25% from the previous year. Bing's advertising contributed most of the revenue. By comparison, Google's search generated at least ten times more revenue than Bing in the same period. In 2021, the advertising business brought in $208 billion for Google, accounting for 81% of Alphabet's total revenue.

But now, the situation has changed. The combination of Bing and ChatGPT has pushed Google to the forefront. Facing a new search revolution, the later the response, the more users may flow to Microsoft Bing, and more personalized customized answers. The core of traditional search engines is to retrieve and aggregate information from massive sources rather than to create information. However, the Bing + ChatGPT combination is not like this; this new "AIGC" product has become an industry innovation.

In addition to search, Microsoft has also updated the Edge browser, promoting ChatGPT-powered Bing to other browsers. Of course, this will also intensify the competition between Microsoft's Edge and Google's browsers in providing better searches, more comprehensive answers, a new chat experience, and content generation capabilities. Microsoft says that in the new version of the Edge browser, Bing's AI features can also present financial results or summaries of other web pages, aimed at allowing readers to avoid understanding lengthy or complex documents, and it also corrects computer code.

It's about more than just market share in search. For tech companies, data is life, just as hardware manufacturers want to get a share of the mobile phone market. Undoubtedly, search engines are the lifeblood that connects every user and generates massive amounts of data and information. It can be said that ChatGPT has stirred up more than just a simple search engine battle; it's a massive data and information battle. And now, Microsoft has taken the lead.

3.2.2 The Forefront of the AI Revolution

Microsoft's integration of ChatGPT into the Bing search engine is considered a major move by Microsoft. In addition, shortly after the release of GPT-4 in March, Microsoft announced the official incorporation of the GPT-4 model into its Office suite, launching the new AI feature, Copilot. In Microsoft's new Copilot system, GPT-4 will be responsible for the mutual invocation of office software such as Word, Excel, PPT, and the Microsoft Graph API.

If the release of the GPT-4 model has already made us aware of the current strength of AI, Microsoft's integration of GPT-4 into the Office suite gives us a more specific understanding of the upcoming AI era. As Microsoft Vice President Jared Spataro said at the start of the launch event, "In a hundred years, we will look back on this moment and say, 'that was the beginning of the true digital age.'"

The integration of GPT-4 into Microsoft's Office suite is truly stunning and can even be considered a milestone in the development of AI. This means how we interact with computers has entered a new stage, opening up an era of AI collaboration with human work.

For example, if we want to write a story in Word, we can give Word a brief description, and it can generate a draft for us. The most powerful thing is that we can input other files directly and specify that the AI creates content based on their content. The generated content is well-organized, and the format is arranged for us.

For all the content generated by Copilot, if we think it's good enough, we can keep it. If unsatisfied, we can adjust the AI settings or try again.

This draft can save time and directly revise and create it. The intelligence of Copilot in Word far exceeds our imagination because it also supports switching between various tones, such as using professional terminology on professional occasions and another description on casual occasions.

In Excel, using Copilot can make it easier to create complex spreadsheets. For users who don't understand various function calls, macros, and VBA language in Excel, based on Copilot, they can directly ask various questions in "human language." It will recommend some practical formulas. Copilot in Excel can also find the correlation of data, generate models based on questions, and draw trends. It also instantly creates data-based SWOT analyses or pivot tables.

A PPT can also be generated directly through Copilot. We only need to enter the information we want to present and the style we want, and then click "generate," and a beautifully formatted and animated PPT is created.

In addition to the Office suite, Microsoft has also embedded Copilot functionality in other software and low-code platforms. In Microsoft Teams, the Copilot feature can transcribe meetings, such as creating a summary of the content from the beginning to the end of the meeting, and can also answer specific questions about the meeting. In addition, Copilot can also generate a meeting agenda based on chat records, suggest who should follow up on specific projects, and suggest a time to schedule attendance.

In Outlook, Copilot can use AI to read emails and then generate replies for you. Business Chat is a new experience released by Microsoft. It uses Microsoft Graphs and AI to collect information from programs such as Word, PPT, email, calendar, notes, contacts, etc. It integrates them into a single chat interface in Microsoft Teams. This interface generates summaries and plans overviews.

For Microsoft, the significance of Copilot is not limited to the traditional Office suite of software but rather encompasses the entire Microsoft Office ecosystem, including email, contacts, online meetings, calendars, and workgroup chats. All data is integrated into the LLM to create the new Copilot system. During an online meeting, Copilot provides real-time summaries and indicates which problems have yet to be resolved and need further discussion.

Microsoft stated at the launch event that they have also released an enterprise-level AI assistant in addition to individual users. Currently, Microsoft is testing Microsoft 365 Copilot functionality with a small number of clients. It plans to release a preview version of Copilot and more pricing information to users in the next few months.

It is foreseeable that Microsoft will incorporate many AI applications into the new Windows 12 operating system, which is expected to be released in 2024. This will completely overturn the series of operating systems before Win11 and become an operating system based on AGI. As the world continues to be changed by AI, this collaboration between Microsoft and OpenAI may just be the beginning. The future is bright, and both Microsoft and OpenAI hope to be at the forefront of this AI revolution.

For us, or workers, the release of Copilot shows us that the iPhone moment for AI is finally here. Copilot is the super application that AI has been waiting for, fundamentally changing how people work by freeing them from repetitive tasks and allowing them to devote more time and energy to more creative work. All we need to do is open Office and tell it our ideas, and Copilot does the rest of the work for us.

According to a survey of developers using Copilot on GitHub, 88% of people reported increased work efficiency, 74% said they could focus on more satisfying work, and 77% said it helped them spend less time searching for information or examples.

From the introduction of AI represented by ChatGPT to the release of GPT-4 to its integration into office software applications, the speed at which AI has entered our human society for collaboration is even faster than we humans expected because every integration exceeds our expectations. Therefore, from the integration of Copilot, it is clear that AI will replace all rule-based and regular work in human society even faster than we expected.

3.3 Google's Response

In 2022, Google, with a market value of $1.4 trillion, generated $163 billion in revenue from its search business. Google has operated for more than 20 years and has maintained a market share of up to 91% in the search field—until the emergence of ChatGPT. Many competitors have tried to compete head-on with Google, but they have all failed. However, at the end of 2022, OpenAI's ChatGPT appeared out of nowhere, causing the search giant Google to sound the "code red" alarm and urgently launch a product to compete with ChatGPT.

What impact has ChatGPT had on Google? How will the temporarily lagging Google respond to this unexpected battle?

3.3.1 The Challenged Search Engine

Google search was once considered an impeccable and irreplaceable product—its revenue and finance were awe-inspiring, it held a leading market position, and users recognized it.

This is, of course, due to the technology behind Google search. The working principle of Google search technology is to use algorithms and systems to index and rank billions of web pages and other information on the Internet and provide relevant results to users in response to their search queries.

Regarding crawling and indexing, Google uses automated bots to scan the Internet and find new or updated web pages. The information from each page is stored in Google's index, a massive database containing information on billions of web pages. Regarding relevance, when a user performs a search, Google uses a set of algorithms to determine the relevance of each web page in its index to the user's query. Relevance is determined by factors such as looking at page content, user location and search history, and the relevance of other pages linked to that page.

In addition, based on the relevance of each page, Google calculates a "rank" for each page and uses it to determine the order in which the pages appear in search results. The most important part of the ranking calculation is the PageRank algorithm, which assigns a rank to each page based on the number and quality of links pointing to that page.

Traditional search engines often check the frequency of keywords appearing on web pages. PageRank technology treats the entire Internet as a whole, examines the structure of the entire network of links, and determines which web pages are most important. Specifically, page A is relatively important if any website links point to page A. PageRank statistically weighs the number of links. For links from important websites, their weight is also greater. This algorithm is entirely without human intervention; vendors cannot buy page rankings with money. Finally, based on the calculated ranking, Google generates a list of relevant results and presents them to users as a search results page. The results are sorted by relevance, with the most relevant results appearing at the top.

The PageRank algorithm allows Google to provide better and more accurate results than its competitors, demonstrating Google's technical strength and playing a big role in Google's growth. This sensitive technology at the time brought excellent products— Google is not only the most useful search engine, but it is also fast and intuitive. For example, other search engines allow advertisers to use images in their postings, but Google does not. It's simple, images slow down web page loading speeds and reduce user experience.

With a good user experience, Google has soared in the search engine industry and maintained a market share of up to 91% in the search field—until ChatGPT appeared.

ChatGPT has transformed search engines from merely search engines to more intelligent and personalized products. Using ChatGPT feels like inputting our requirements into a smart box and receiving a well-crafted written response. This response is not only free from the influence of images, advertisements, and other links, but it also "thinks" and generates content that it believes can answer your questions. This is evidently more appealing than traditional search engines.

3.3.2 Google's Response to ChatGPT

ChatGPT has attracted worldwide attention and made Google feel the crisis.

Regarding investment, on February 4, Google Cloud, a subsidiary of Google, announced that it had established a new partnership with OpenAI's competitor, Anthropic, and Anthropic has chosen Google Cloud as its preferred cloud provider to provide the computing power needed for its AI technology. According to the *Financial Times*, Google has invested about $300 million in Anthropic, acquiring 10% of the company, and the new financing will increase Anthropic's post-investment valuation to nearly $5 billion.

In terms of products, on February 7, Google CEO Sundar Pichai announced the launch of a conversational AI service supported by the LaMDA model called Bard. Pichai called it the "important next step in Google's journey in AI." In a blog post, he introduced Bard as seeking to combine the breadth of world knowledge with the power, wisdom, and creativity of LLMs. It will use information from the web to provide fresh, high-quality responses. It is both an outlet for creativity and a launcher of curiosity. He also stated that the eligibility to use Bard would be first given to "trusted testers" and then opened to a wider public in the coming weeks.

However, ChatGPT has already deeply bound with Microsoft, and ChatGPT's API will also be placed on Microsoft's Azure cloud. Google is almost the only serious competitor in this field. Although not naming names, Bard's conversational AI service is obviously a competitive product launched by Google to respond to OpenAI's

ChatGPT, and adding more powerful AI functions to the search engine is also to counter Microsoft's Bing search engine under the support of ChatGPT.

Unfortunately, when Google released Bard, it made a factual error in the first online demo video. In an animation shared by Google, Bard answered a question about James Webb Space Telescope's discovery, stating that it "took the first batch of photos of planets outside the solar system." However, this is incorrect. The VLA in Chile took the first picture of a planet outside our solar system, i.e., an exoplanet, in 2004. An astronomer pointed out that this issue might have arisen due to the AI's misunderstanding of a "vague news release from NASA." This mistake also caused a plunge of about 8% in the opening price, evaporating $102 billion (¥693.25 billion) in market value. Bard's mistake caused Google's market value to evaporate by billions, and the official launch date of Bard is still uncertain.

Earlier, Pichai also pointed out that Bard "uses web information to provide fresh, high-quality responses," indicating that it may be able to answer questions about recent events that ChatGPT finds difficult to solve. For example, Bard can help you explain NASA's James Webb Space Telescope's discovery to a nine-year-old child or provide information about the best striker in the current football world. It is believed that this hastily and unclear announcement about Bard is likely a result of Google's "code red." Previously, Google's subsidiary, DeepMind, released the chatbot Sparrow, and Google also launched the AI music model MusicLM.

During a Q&A session on February 3, Pichai said that Google will launch a LLM based on AI like ChatGPT in the "coming weeks or months," and users will soon be able to use the language model as a "search companion."

Paul Buchheit, Google's 23rd employee and creator of Gmail, previously stated that Google will be completely disrupted within one or two years. When people's search needs can be met with packaged, semantically clear answers, search ads will have no survival value. Despite holding nearly 84% of the global search market, Google remains a company where 50% of its revenue comes directly from search ads.

3.3.3 Google's Crisis about ChatGPT

In fact, Google's achievements in the AI field are not inferior to any other tech giant. In 2014, Google acquired DeepMind, which was considered a win-win situation. On the one hand, Google acquired the industry's top AI research institution, while the money-burning DeepMind also received strong funding and resource support. DeepMind has always been Google's pride. As a subsidiary of Alphabet, DeepMind is one of the world's leading AI laboratories. In its 13-year history, it has delivered impressive results.

In 2016, the program AlphaGo, developed by DeepMind, challenged and defeated the world Go champion Lee Sedol, and the related paper also appeared on the cover of the journal *Nature*. Many experts believe that this achievement came decades earlier than expected. AlphaGo demonstrated creative ways of winning games and, in some cases, even found moves that challenged thousands of years of Go wisdom.

In 2020, AlphaFold made a huge splash. After the success of the AlphaGo algorithm in Go games, DeepMind turned to protein structure prediction based on amino acid sequences, proposed a deep learning algorithm called AlphaFold, and achieved outstanding results in the international protein structure prediction competition CASP13. DeepMind also plans to release more than 100 million structure predictions, equivalent to nearly half of all known proteins, hundreds of times the number of experimentally solved proteins in the Protein Data Bank structure database.

Over the past half-century, humans have analyzed the structures of more than 50,000 human proteins, and about 17% of the amino acids in the human proteome have structural information. AlphaFold's predicted structure raises this number from 17% to 58%, which is almost the limit since the proportion of amino acids without a fixed structure is large. This is a typical case of a quantitative change leading to a qualitative change in just one year.

In October 2022, AlphaTensor, developed by DeepMind, appeared on the cover of *Nature*, becoming the first AI system to discover novel, efficient, and correct algorithms for basic computing tasks such as matrix multiplication.

In addition, Google's invention of Transformer is a key technology that supports the latest AI models and is also the underlying technology of ChatGPT. The initial Transformer model had a total of 65 million tunable parameters. The Google Brain team used various publicly available language datasets to train this initial Transformer model, including the 2014 English-German Machine Translation Workshop (WMT) dataset (with 4.5 million pairs of English-German sentences), the 2014 English-French Machine Translation Workshop dataset (with 36 million pairs of English-French sentences), and some sentence pairs from the Penn Treebank language dataset (including 40,000 sentences from the *Wall Street Journal* and an additional 17 million sentences). Moreover, the Google Brain team provided the model's architecture in the paper, so anyone can build a similar model and train it with their data. In other words, the AI Transformer model built by Google is an open-source or open-source underlying model.

At that time, the initial Transformer model launched by Google achieved industry-leading scores in translation accuracy, English constituent syntactic analysis, and other evaluations, becoming the most advanced large-scale language model at the time. ChatGPT uses the technology and ideas of the Transformer model and expands and improves it based on the Transformer model to better apply it to language generation tasks. Based on the Transformer model, ChatGPT has achieved its success today.

Moreover, Google also had the opportunity to follow the path of ChatGPT. In the field of chatbots, Google is not at a disadvantage. At the I/O conference in May 2021, Google may still be a leader in the field of LLMs and a force to be reckoned with in terms of innovative AIGC, as Emad Mostaque, founder of Stability AI, commented on Google's crisis. However, it is clear that the rise of startups like ChatGPT and the success of their AI language models threaten Google's dominance.

In the end, it remains to be seen how Google will respond to the challenge posed by ChatGPT and other startups. Will they continue focusing on their core search and ad revenue business, or will they try to catch up with the AI language model trend and develop their cutting-edge models? Only time will tell, but one thing is for sure: the era of AI language models has just begun and is set to revolutionize how we interact with technology and each other.

In addition, Google also had the opportunity to follow the ChatGPT route. In the field of chatbots, Google is staying caught up. As early as Google's I/O conference in May 2021, the company's AI system, LaMDA, impressed everyone with its ability to make answers to questions more natural. Even last June, Google engineer Blake Lemoine had an emotional conversation with LaMDA and believed that it had the intelligence of an eight-year-old child and was also "conscious." According to some leaks, Google's LaMDA chatbot far outperforms ChatGPT's performance.

Moreover, Google claims its Imagen model's image generation ability is superior to DALL·E and other company models. However, it is somewhat embarrassing that Google's chatbot and image model only exists in the "claimed" stage, and there are no actual products on the market.

It is easy to understand why Google is doing this. On the one hand, Google has always believed that using ML to improve search engines and other consumer-facing products and providing Google Cloud technology as a service is its core business. The search engine has always been Google's core business.

On the other hand, Google is concerned that immature AI chatbots may make some ridiculous mistakes, which could lead to "reputation risks" for the company. At the same time, Google is also concerned that chatbots that can accurately answer users' questions may subvert the company's current core business, search engines, and erode the company's profitable advertising business.

However, Google did not anticipate that LLMs like ChatGPT would bring disruptive innovation to business. As disruptive products become better and better, they naturally pose more and more severe challenges to Google. In this case, Google's recent move was to upgrade its search engine, allowing users to enter fewer keywords and get more results. Compared with Google, the ChatGPT startup team has no historical business burdens and pursues relatively pure goals to achieve human-like levels in text processing and communication through AI. They seek a subversion that makes technology and products truly usable and intelligent.

Regarding the crisis facing Google, Emad Mostaque, the founder of Stability AI, commented that Google is still a leader in the field of LLM. In the innovation of AIGC, they are a force that cannot be ignored. On March 29, 2023, The Information revealed the shocking news that Google's Bard was actually trained on ChatGPT data, disclosed by Google's ex-employee who had jumped ship to become a top researcher at OpenAI, Jacob Devlin. This shows that challenging ChatGPT is not an easy task, and Google is clearly facing increasing pressure, and being dismantled may just be a matter of time.

3.4 Apple's Response

After Microsoft embraced ChatGPT, other tech giants took action, with Apple CEO Tim Cook stating during the February 3 earnings call that AI is a key focus for Apple. Cook emphasized that AI is a horizontal technology that will impact all Apple products and services.

3.4.1 *Infinite Possibilities of AIGC*

The emergence of ChatGPT has both positive and negative implications for Apple. On the one hand, ChatGPT has shown the unlimited potential of AIGC, and Apple has already been developing its own AI for text and image generation, such as Stable

Diffusion, which has been optimized to run on consumer-grade GPUs and even on iPhones.

In December 2022, Apple's ML team announced that they had optimized the Stable Diffusion model and updated their operating system to accommodate it for their chips. Stable Diffusion can also be embedded into the Apple operating system and provide easy access to its API for developers. This move puts Apple in a strong position with its ecosystem advantage, while independent app makers have access to APIs and distribution channels to build new businesses.

Although Stable Diffusion on Apple devices may not dominate the entire market, its embedded local capabilities will influence the final processing market for centralized services and computing.

3.4.2 *The Challenge of User Stickiness*

On the other hand, the impact of ChatGPT, combined with Google's release of its AI chatbot Bard and Microsoft's upcoming AI event, has also brought significant challenges and pressures to tech giants like Apple. Apple seems to lag behind its competitors, such as Google, Facebook, Microsoft, and Amazon, in AI & ML, mainly reflected in Siri, the intelligent voice assistant product embedded in the iPhone. Although Siri is the most direct manifestation of Apple's AI technology productization and has been optimized for years, it is often criticized by users and lacks sufficient user stickiness.

When a technology lacks sufficient user stickiness, the most immediate problem is that the product's user experience is significantly different from what users expect or the product experience is not good. Poor user experience further reduces user enthusiasm, reducing the acquisition of training data. This will result in a vicious cycle of AI product iteration, slowing down the product's upgrade optimization speed, which is precisely the predicament currently facing Apple's Siri.

Therefore, some in the industry believe that Apple is a "latecomer" in the field of AI. Cook also recognized the importance of AI for a giant company in this era and is vigorously promoting Apple's research in AI. Cook also recognized the gap between Apple and other giant technology companies in the field of AI and is now trying to correct this gap, especially in the current AI revolution sparked by ChatGPT, which is particularly important for Apple.

According to Mark Gurman, influenced by recent AI technology events, Apple will hold its annual internal AI summit in February at its headquarters. This AI summit is like a WWDC conference specifically for AI but limited to Apple employees only. Perhaps we also understand this as Apple's anxiety about coping with the AI revolution sparked by ChatGPT and holding an internal conference to obtain better suggestions and solutions.

Of course, it is not to say that Apple has made no progress in AI, but most of its technology is deeply bound to its products. Its exploration direction in AI is more focused on using AI to improve its product performance, such as chips and photography, image and classification, and is also a "hidden" approach. Apple's slow iteration and upgrade strategy, whether in software or hardware, was a perfect strategy before the revolutionary change sparked by ChatGPT and is a strategy that maximizes business interests. However, without ChatGPT-like performance products, Apple will face a significant crisis. This technology is precisely what Apple's Siri has been dreaming of. Whether for Siri or the IOS ecosystem, if Apple continues to focus its AI research on the optimization level of service hardware, it will face enormous risks. Although we are still unclear about what kind of cards Apple has in AI after so many years, time is running out for Apple. It is believed that this year, Apple's AI technology will be revealed to everyone.

Of course, there is also an option for Apple to choose to cooperate with OpenAI. However, the cost of such cooperation for Apple may be enormous, and its closed ecosystem will be challenged.

3.5 Meta's Challenges in Exploring AI

Currently, Meta is accelerating the commercialization of AI, and AIGC is undoubtedly a huge opportunity for Meta. However, it is still unknown whether Meta's business empire will exist until the day when the metaverse strategy is realized.

3.5.1 Evaluations of ChatGPT

It is worth mentioning that Yann LeCun, the chief AI scientist at Meta, does not hold a high opinion of ChatGPT regarding its underlying technology. He believes that from a technical perspective, ChatGPT is not an innovative or revolutionary invention. LeCun stated that many companies and research laboratories have built this kind of data-driven AI system in the past, and the idea that OpenAI is a lone warrior in this work needs to be more accurate. In addition to Google and Meta, several startups have similar technologies. Furthermore, LeCun also pointed out that ChatGPT and its underlying GPT-3 comprise multiple technologies developed by multiple parties over many years. LeCun believes that instead of saying ChatGPT is a scientific breakthrough, it is a decent engineering example.

LeCun's comments have sparked controversy. Some criticize him for being too arrogant, picking holes in widely available products he has not developed. Regardless of whether it can be called a "breakthrough," the success of ChatGPT is undeniable. Therefore, people cannot help but wonder whether Meta's AI team FAIR (Facebook AI Research), led by LeCun, will achieve breakthroughs in the public's eyes like OpenAI.

LeCun's answer is affirmative. LeCun has set the goal of "generative art" for FAIR. He proposed that there are 12 million stores advertising on Facebook, many of which are small shops without resources for customized advertising. Meta will help them do better promotion by using AI that can automatically generate promotional materials. When asked why Google and Meta have not launched a ChatGPT-like system, LeCun replied, "Because Google and Meta will suffer huge losses from launching a system that fabricates things." In contrast, OpenAI seems to have nothing to lose.

In fact, Yann LeCun's evaluation of ChatGPT is a common view among some AI experts in the field, which is to view ChatGPT from a professional technical perspective rather than a business application perspective. Just like Google, which has the fastest pace of development, the broadest research, and the strongest technical strength in AI it still lags behind OpenAI in the phenomenal application of AI and faces enormous challenges.

3.5.2 The Difficulty to Implement AI

In fact, Meta has been investing heavily in AI research since 2013, and FAIR, led by Yann LeCun, has been at the forefront of the era alongside DeepMind and OpenAI for a long time. In January 2022, FAIR was merged into the Reality Lab as a subdepartment.

In 2022, Meta's progress in the generation model has also been accelerating. In January 2022, it released the speech generation model data2vec, which can learn speech, vision, and text in the same way, and released data2vec 2.0 in 2022, which greatly improves its training and inference speed. In May 2022, it released the open-source language generation model Open Pretrained Transformer (OPT), which uses 175 billion parameters like GPT-3. Then in December 2022, it released its updated version OPT-IML, which will also be free for non-commercial research purposes. In July 2022, it released the image generation model Make-A-Scene. And in September 2022, it released the video generation model Make-A-Video.

However, Meta's exploration of ChatGPT has also been hindered. On November 15, 2022, Meta's AI language model Galactica, which was developed to use ML to "sort scientific information," briefly went online and spread a large amount of false information. Within 48 hours, Meta's AI team hastily "paused" the demonstration. Media analysis suggests that while Meta stumbled due to Galactica's failure, other Silicon Valley Internet giants have poured huge sums of money into the "AIGC" craze, leaving Meta struggling to catch up.

Objectively speaking, Meta's advantage lies in its huge data centers, but mainly CPU clusters, which support Meta's deterministic advertising model and network content recommendation algorithm business.

However, Apple's transparency tracking technology has dealt a major blow to Meta. On April 26, 2021, Apple's App Tracking Transparency (ATT) privacy collection permit policy was officially implemented, giving users the right to choose whether to be tracked by app developers, thus blurring the attribution of advertising effectiveness. Before ATT, Meta was the largest platform for advertising placement in the US market. It could collect data from advertising clients' internal applications and websites, clarifying which advertisements led to which results. This gave advertisers the confidence to spend money on advertising, not caring about cost input but focusing on how much revenue they could generate. ATT severed the link between Meta's advertising and conversion, marking the latter as third-party data and therefore tracking it. This reduced the value of the company's advertising and increased the uncertainty of advertising conversion. On Apple's policy announcement day, Facebook's stock price fell by 4.6%. It can be said that ATT has done more damage to Meta than any other company.

Although the long-term solution to ATT is to establish probabilistic models, which need to understand customer goals and which advertisements have converted and which have not, these probabilistic models will be built by large-scale GPU data centers. One NVIDIA graphics card costs five figures, and if it's a deterministic advertising model like before, Meta doesn't need to invest more in GPUs. Still, technology is advancing, and Meta needs to face a new era and invest more in customer targeting and conversion rates.

An important factor in allowing Meta's AI to function is not simply building basic models but constantly adjusting them for individual users. This is the most complex part, and Meta must figure out how to provide personalized user services at low cost while its products become increasingly integrated. Obviously, recommending content from the Internet is much more difficult than recommending content from your friends and family, especially since Meta plans to recommend videos and all types of media and mix them with the content you care about. AI models will be critical and building these models will require significant capital investment in purchasing equipment. In the long run, although Meta's previous investment in AI focused on personalized recommendations, combining these investments with the breakthroughs in generating models in 2022 will ultimately result in personalized content, which will be delivered through Meta's channels. As Sam Lessin once said, the ultimate goal of algorithms is AIGC.

However, a huge dilemma for Meta on its path to AIGC is knowledge pollution and data governance. Since humans gained access to the Internet, the amount of data produced has grown exponentially. In the field of Internet socialization, data chaos is the highest in the entire Internet industry, especially since entering the AIGC era from the UGC era. The self-replicating ability of non-essential factual data will exceed human imagination and be difficult to control. The human knowledge database will inevitably encounter a pollution crisis, and AIGC trained based on this data will eventually slide into the abyss of data loss and chaos.

Building a future on the metaverse may be a long and uncertain road because all the cutting-edge technologies required, except for AI, have yet to make fundamental breakthroughs.

3.6 Amazon's Response

3.6.1 The Converging of AI and Cloud Computing

In the face of the explosion of ChatGPT, Amazon's advantage in C-end scenarios such as image and text generation seem not so obvious. However, what is important is AWS—the pioneer and leader in cloud computing. Its success ignited the cloud computing revolution, a new opportunity for Amazon in the ChatGPT wave.

Since 2022, AIGC has successfully made breakthroughs in image generation, and ChatGPT has brought hope for the entire AI field to leap into the 2.0 stage. Among them, Stability AI's image generation engine is driven by its open-source algorithm Stable Diffusion. Its keyword-generated images won awards in competitions and made graphic designers feel unprecedented competitive pressure. During the training phase, Stable Diffusion ran on 150,000 GPUs. After commercialization, Stability AI quickly cooperated with AWS to build a large-scale cloud computing cluster consisting of 4,000 A100 GPUs.

With its powerful computing resources, Stability AI is preparing to enter the next hot field of AI for Science. It has gathered well-known open-source projects such as EleutherAI and LAION and cutting-edge explorations such as OpenBioML for biological modeling, Harmonai for audio generation, and Carperai for human preference learning. The future use of diffusion models to generate DNA sequences is a research direction that may benefit billions of people worldwide. The success of Stability AI is not accidental. The high integration of AI and cloud computing promotes the rapid landing of various applications. In the past, various applications were migrated to the cloud. Still, the applications themselves are evolving toward cloud-native, which has become a consensus for more forward-looking technology giants. Cloud-native applications train algorithms in the cloud and integrate the entire development, delivery, deployment, and operation process in the cloud. Using a cloud-native development environment, AI can significantly reduce the cost of configuring servers and save the transmission cost of massive training data.

According to the latest financial reports, top global cloud providers such as Amazon, Microsoft, and Google have all experienced a slowdown in revenue growth in their cloud computing businesses. This makes them eager to find breakthroughs and open up incremental markets. The global explosion of ChatGPT sends a signal: the potential prosperity of AIGC may once again boost market demand for cloud services.

Microsoft management said in a conference call after the 2023 Q2 (i.e., Q4 2022) earnings report that Microsoft is revolutionizing the computing platform with AI models, and a new wave of cloud computing is emerging. Google has also joined the AI computing competition. Google CEO Sundar Pichai said, "Our latest AI technologies, such as LaMDA, PaLM, Imagen, and MusicLM, are creating new ways to process information, from language and images to video and audio." Obviously, Google and Microsoft have the same attitude and regard AI computing as a focus of competition for dominance.

As a typical platform enterprise, Amazon focuses on providing basic public cloud services to users, such as computing, storage, networking, and databases, and rarely touches upper-level applications, leaving space for partners. In fact, AWS is Amazon's most significant commercial competitive product, and it has grown into the world's largest public cloud platform. At the infrastructure level, AWS has 87 available zones across 27 global geographic regions, covering 245 countries. AWS accounts for over one-third of the global public cloud market in terms of market share. In terms of product services, AWS is the most comprehensive cloud platform globally, offering more than 200 services with comprehensive functions, and the number of new functions or services launched each year is rapidly increasing. At the user and ecosystem level, AWS has built specialized teams for key industries such as finance, manufacturing, automotive, retail and consumer goods, healthcare and life sciences, education, gaming, media and entertainment, e-commerce, and energy and power. This has enabled Amazon to have millions of customers and the largest and most vibrant community.

Amazon holds ChatGPT in high regard. Amazon has used ChatGPT in many job functions, including answering interview questions, writing software code, and creating training documents. However, Amazon needs to pay attention to the AI computing market.

In fact, Amazon AWS offers a variety of AI services, including Amazon Lex, Amazon Rekognition, and Amazon SageMaker, covering areas such as speech synthesis, natural language generation, and CV. As with the collaboration between Stability AI and AWS, Amazon SageMaker, a heavyweight product from AWS, can easily deploy pretrained models in a browser, and the subsequent fine-tuning and secondary development processes save on tedious configuration.

Moreover, AI computing requires large-scale procurement of GPU computing power. According to Emad Mostaque, the founder and CEO of Stability AI, the company used 256 NVIDIA A100 GPUs for 150,000 hours of usage at a market price of $600,000. However, the greater demand is for inference, the actual application of models to generate images or text. Each time an image is generated in Midjourney or a profile picture is generated in Lensa; the inference runs on cloud GPUs.

Currently, the NVIDIA A100 GPU with 80GB of memory is priced at around $17,000, and each card costs about $4/hour to rent on a cloud computing platform. Stable Diffusion requires 256 A100 GPUs for training, taking approximately 24 days, and pays AWS for 150,000 hours. However, this is already a low price compared to language generation models that cost millions or even tens of millions of dollars with hundreds of billions of parameters.

Of course, in AI computing, Amazon still needs to face the impact of the binding between ChatGPT and Microsoft on its business. At present, ChatGPT's parent company, OpenAI, has not only established its model but also reached a preferential agreement with Microsoft for computing power. In the long run, Amazon may have to sell GPU computing power at lower prices to stimulate more prosperous generative applications.

3.6.2 Challenges to Amazon's E-commerce

For a long time, Amazon has been the "king of e-commerce" globally, and this is due to several killer features it has built over the years. In summary, these are divided into three main points: first, the FBA (Fulfillment by Amazon) service on the seller side; second, the Prime membership model on the customer side; and third, Amazon's technical support.

In terms of technology, in addition to cloud technology, Amazon's ML has a history of over 20 years. As early as 1998, Amazon.com launched an item-based collaborative filtering algorithm, which was the first application of a recommendation system on a million-item and million-user scale. For example, the "customers who bought this also bought" feature on Amazon's website is supported by collaborative filtering algorithms.

This feature alerts users in the marketplace by saying, "Other customers who bought this item also bought." The algorithm recommends items based on "similar purchasing behavior of users who may like the same items." The algorithm first evaluates the similarity between users based on their purchase history and then recommends products based on the preferences of other users.

Other ideas of collaborative filtering algorithms include "similar items may be liked by the same user" (such as recommending basketballs to users who have purchased basketball shoes) and model-based collaborative filtering (such as SVD), which is used to cope with the high computational complexity resulting from the massive data scale of Amazon's marketplace. This technology has led to the innovative "thousand faces of Amazon" personalized recommendation system that is renowned in the industry.

Personalized recommendations can increase content interaction, reduce customer acquisition costs, and improve user retention and stickiness. Improving customer acquisition opportunities improves overall business efficiency, enabling deeper demand mining of users, such as in promotion scenarios. Suppose users cannot be mined at a deeper level. In that case, it is difficult to achieve a thousand faces, making it difficult to penetrate long-tail products and negatively impacting long-term operations.

In this context, in March 2021, Amazon launched Amazon Personalize, a fully managed ML service for building personalized recommendation systems. With over 20 years of experience in personalized recommendations, Amazon Personalize is an attempt to turn Amazon's accumulated recommendation technology into a platform for external services.

On the one hand, the recommendation filters in Amazon Personalize can help users fine-tune the recommended content based on business needs without the need to design any post-processing logic. The recommendation filters filter and recommend products that users have purchased, videos they have watched in the past, and other digital content they have consumed, thereby improving the accuracy of personalized recommendations. In the past, recommended content provided by recommendation systems often needed higher accuracy, which could affect user emotions, reduce user engagement, and ultimately lead to revenue loss.

On the other hand, when new users enter an e-commerce website, the website immediately predicts their potential shopping needs based on some basic registration information. Amazon Personalize effectively generates recommendations even for new users and finds relevant new item recommendations.

It can be said that Amazon's "personalized recommendations" are the foundation of Amazon's e-commerce business. According to Amazon's financial reports in recent quarters, the company's advertising revenue is steadily increasing, and these revenues are obtained from sellers. In 2014, the cost per click for advertising on Amazon was only about 0.14 cents, but by early 2022, it had soared to about $1.60 per click.

In this context, Amazon's dominance in the e-commerce space has been challenged by the emergence of new independent e-commerce platforms such as Shein, Shopify, and Temu. These platforms are gradually breaking through Amazon's e-commerce stronghold. In contrast to Amazon's FBA service for third-party sellers, Shein's model allows sellers to be more relaxed and "lie flat."

In 2006, Amazon introduced FBA, a one-stop fulfillment service for third-party sellers. FBA charges sellers for storage, sorting, distribution, customer service, and returns. This reduces the seller's time and operating costs and attracts many merchants to participate. However, under the Amazon model, in addition to paying for FBA warehousing and transportation costs, sellers also need to bear the cost of initial transportation, product commissions, and product promotion work and costs. In Shein's model, sellers only need to stock and wait for collection, and the platform handles all logistics, sales promotion, operation management, and return and exchange services. If Shein can leverage ChatGPT to provide personalized, low-barrier recommendations based on user needs, adapt to the ever-changing user business models and environments, and achieve "change on demand," Amazon's e-commerce business will face unprecedented challenges.

3.7 NVIDIA's Achievements

In the current technology industry where silence is golden, ChatGPT may be the most dazzling presence, so much so that even Bill Gates believes that this type of AI technology is of no less historical significance than the birth of the Internet and personal computers. As ChatGPT surges ahead and ignites the multi-billion-dollar AIGC race, another tech giant is quietly raking in cash—NVIDIA.

On January 3, 2023, the first trading day of the US stock market, NVIDIA's closing price was $143. One month later, on February 3, NVIDIA's stock price rose 47% to $211. Wall Street analysts predict that NVIDIA's performance in January alone will add $5.1 billion to the personal assets of its founder, Jensen Huang. According to the "Bloomberg Billionaires Index," Huang has become the person with the largest increase in personal wealth among American billionaires this year.

The ups and downs of semiconductor stock prices are normal, but today is different. The semiconductor market is experiencing a rare downward cycle. ChatGPT's success has led to a significant increase in NVIDIA's stock price because NVIDIA's hardware support is integral to ChatGPT's success.

3.7.1 *The First AI Chip Stock*

In the 1990s, the rapid development of 3D games and the gradual popularity of personal computers completely changed the logic and creative process of gaming. In 1993, Jensen Huang and three other electrical engineers saw the demand for 3D graphics processing power in the gaming market and founded NVIDIA to supply graphics processors to the gaming market. In 1999, NVIDIA launched the graphics card GeForce 256 and, for the first time, defined the graphics processor as "GPU," a term and standard that NVIDIA popularized in the gaming world.

Since the 1950s, the CPU has been the core of every computer or intelligent device and the only programmable component in most computers. Even after the birth of CPUs, engineers have always kept their efforts to make CPUs achieve the fastest calculation speed with the least energy consumption. However, people have found that CPUs are too slow for graphics computing. In the early 21st century, CPUs could not maintain a 50% performance improvement each year. GPUs, which contains thousands of cores, were able to continue to improve performance by utilizing their inherent parallelism. Their massively parallel architecture is more suitable for high-concurrency deep learning tasks.

Compared to CPUs designed to complete tasks as quickly as possible to reduce system latency and switch between different tasks quickly to ensure real-time performance, most CPUs also execute tasks serially. In contrast, GPUs are designed to increase system throughput and simultaneously process as many tasks as possible. This characteristic has gradually caught the attention of developers in deep learning. However, as a graphics processing chip, GPUs are difficult to develop in the general computing field, like CPUs using high-level programming languages like C and Java, which greatly limits their development.

NVIDIA quickly noticed this need. In order to allow developers to use NVIDIA GPUs to execute computing tasks beyond graphics processing, NVIDIA launched the CUDA (Compute Unified Device Architecture) platform in 2006, which supports developers to use familiar high-level programming languages to develop deep learning models, flexibly call NVIDIA GPU computing power, and provide databases, debugging programs, API interfaces and a series of tools. Although deep learning, which was thriving then, did not bring significant benefits to NVIDIA, NVIDIA continued to invest in the CUDA product line and promoted the development of GPUs in AI and other general computing fields.

Six years later, NVIDIA finally got the opportunity to prove GPUs in AI computing. In the 2010s, the "Large Scale Visual Recognition Challenge" held by the large-scale visual database ImageNet project was one of the iconic events in deep learning. It was known as the "Olympics" of CV. In 2010 and 2011, the lowest error rates for the ImageNet challenge were 29.2% and 25.2%, respectively, and some teams had error rates as high as 99%. In 2012, Alex Krizhevsky, a PhD student from the University of Toronto, trained a neural network model with 1.2 million images. Unlike his predecessors, he used NVIDIA GeForce GPUs to provide computing power for training. In that year's ImageNet, Krizhevsky's model won the championship with an error rate of about 15%, shocking the neural network academic community.

This landmark event proved the value of GPU for deep learning and broke the computational limitations of deep learning. Since then, GPUs have been widely used in large-scale concurrent computing scenarios such as AI training.

In 2012, NVIDIA collaborated with the Google AI team to create the largest artificial neural network at the time. By 2016, Facebook, Google, IBM, and Microsoft's deep learning architectures were running on NVIDIA's GPU platform. In 2017, NVIDIA GPUs were introduced into servers by manufacturers such as HP and Dell and used by companies such as Amazon, Microsoft, and Google for cloud services. In 2018, NVIDIA's Tesla GPU, designed for AI and high-performance computing, accelerated the fastest supercomputers in the US, Europe, and Japan.

Its stock price and market capitalization are growing along with NVIDIA's AI territory. In July 2020, NVIDIA's market value surpassed Intel's for the first time, making it the true "first stock of AI chips."

3.7.2 *Selling Water in the Gold Rush*

As we enter 2023, ChatGPT continues to be a phenomenon in the field of AI, and the more popular it becomes, the higher the cost.

The reason for this is that although ChatGPT is capable of conversing by learning and understanding human language, interacting based on context, truly communicating like a human, writing articles, fixing bugs, and analyzing problems dialectically, all of this is achieved through billions of training parameters. As a result, every time ChatGPT answers a user's question, it needs to perform model inference from this vast sea of parameters, which is much more expensive than one might think. After all, if AI products are to become more intelligent, they need to be trained, and computing power is the "energy" that drives AI's gradual intelligence in continuous learning. NVIDIA is currently the "first" in AI computing acceleration, and its Hopper H100, released in April of last year, is also currently the strongest AI GPU.

After more than a decade of technological accumulation, NVIDIA's CUDA ecosystem, which has been developed for GPU's general-purpose computing, including parallel computing platforms and programming models, has become the best choice for efficient computing on large datasets. The CUDA library, tools, and resource ecosystem also makes it easy for developers to take advantage of the parallel computing capabilities of GPUs to build more powerful and efficient AI models. This is also the key to achieving high performance, generality, ease of use, and deep optimization for different application scenarios.

In the gold rush of ChatGPT, NVIDIA is like a "water seller," which is still important and indispensable. According to Guo Junli, IDC's Asia-Pacific Research Director, in terms of computing power, ChatGPT has imported at least 10,000 NVIDIA high-end GPUs, with a total computing power consumption of 3640PF-days. Moreover, ChatGPT will likely drive NVIDIA's related products to achieve sales of $3.5 billion to $10 billion in 12 months.

In fact, before ChatGPT, Stable Diffusion, an AI drawing tool that stirred up the AI graphics computing field, was born after a month of training on a cluster of 4,000 Ampere A100 graphics cards.

Both OpenAI, Microsoft Cloud, and Google Cloud rely on NVIDIA's underlying chip computing power support because the success of ChatGPT is inseparable from it. As a tech giant with a market value of $500 billion, NVIDIA's data center business, represented by the Hopper accelerator card, is its "money printing machine." Credit Suisse analyst Timothy Arcuri said ChatGPT had imported a minimum of 10,000 NVIDIA high-end GPUs to train the model.

Although NVIDIA has not made any official statements about ChatGPT, Citigroup analysts believe that ChatGPT will continue to grow, possibly leading to an increase in sales of the entire NVIDIA GPU chip in 2023, estimated to be between $300 million and $11 billion. Other analysts from Bank of America and Fidelity also believe that NVIDIA will benefit from the popularity of AI and ChatGPT businesses.

3.7.3 *Challenges of NVIDIA*

From a chip perspective, NVIDIA's dominant position is beyond doubt: its market share has remained stable at around 80%. According to data from Top500.Org, the penetration rate of NVIDIA's GPU products in Top 500 supercomputing centers worldwide is increasing yearly. Currently, the demand for AI computing power in the field of AI doubles about every 3.5 months, resulting in a constant shortage of chips. Even though the latest generation H100 chip has been released, the market price of the previous generation A100 chip is still rising compared to the initial release.

Moreover, new products specifically targeting ChatGPT have yet to be seen. It is worth mentioning that ChatGPT, as a star product, has attracted the attention of the whole society to AIGC and large model technology. The trend toward greater demand for chip usage and higher specifications is becoming clear. In the future, large models will become important production tools in the field of AI technology, requiring stronger training and reasoning capabilities to support the efficient completion of computations for massive data models. These requirements will also place higher demands on chip computing power, storage capacity, software stacks, bandwidth, and other technologies.

This also poses a challenge for NVIDIA. On the one hand, when ChatGPT reaches maturity, its computing power base may gradually shift from NVIDIA's dominant position toward a more fragmented market, thereby compressing NVIDIA's profit space in this field. Especially with the outbreak of the AIGC industry represented by ChatGPT, GPUs and new AI chips have gained more possibilities and new opportunities.

From the perspective of language-based generative models, well-distributed computing support is required due to the enormous number of parameters. Therefore, GPU manufacturers with a complete layout in this ecosystem have more advantages. This systemic engineering problem requires a complete software and hardware solution, and in this regard, NVIDIA has launched the Triton solution in combination with its GPU. However, in the case of image-based generative models, although the number of parameters is also large, it is one to two orders of magnitude smaller than that of language-based generative models. In addition, convolutional calculations are still heavily used in their computations. Therefore, in inference applications, AI chips may have some opportunities if they can be well-optimized.

Currently, the design of AI chips is mainly aimed at smaller models, while the demand for generative models is still much greater than the original design goal.

GPUs prioritize flexibility over efficiency in their design, while AI chip design takes the opposite approach, aiming for efficiency in target applications. Therefore, as generative model designs become more stable, AI chip design has the potential to catch up with the iterative development of generative models and even surpass GPUs in terms of efficiency in the generative model field.

On the other hand, the explosive growth of the AIGC industry has placed increasingly higher demands on computing power. However, due to physical process constraints, there are still limitations to improving computing power. In 1965, Intel co-founder Gordon Moore predicted that the number of components that could be accommodated on an integrated circuit would double every 18 to 24 months. Moore's Law summarizes the pace of technological progress in the world. However, physical constraints will eventually limit classical computers that rely on "silicon transistors" as the basic component structure. As transistors get smaller and smaller, the barrier in the middle also becomes thinner. At 3nm, only a dozen or so atoms are blocking the way. At the microscopic level, electrons undergo quantum tunneling, making it difficult to accurately represent a "0" or "1," which is the reason why Moore's Law is said to have hit the ceiling.

Although researchers have proposed replacing materials to enhance the barrier inside the transistor, the fact is that no matter what material is used, it cannot prevent the quantum tunneling effect of electrons. This difficult problem is a natural advantage of quantum computing. After all, semiconductors are products of quantum mechanics, and chips are developed by scientists who understand the quantum properties of electrons. In addition, based on the superposition properties of quantum mechanics, quantum computing is like the "5G" of computing power, bringing speed and other changes.

With its powerful computational capabilities, quantum computing can quickly complete calculations that electronic computers cannot. The growth in computing power brought about by quantum computing may completely break the computing power limitations of current AI large models, promoting another leap in AI.

However, NVIDIA needs a clear advantage in quantum computing compared to Google, which established its quantum computing project in 2006. In October 2019, Google announced in *Nature* that it had achieved quantum supremacy using its 54-qubit processor Sycamore. In addition to Google, in 2015, IBM published a prototype circuit of a quantum chip made of superconducting materials in *Nature Communications*. Intel has been researching multiple qubits, including superconducting and silicon spin qubits. In 2018, Intel successfully designed, manufactured, and delivered the Tangle Lake, a 49-qubit superconducting quantum computing test chip that is equivalent to the computing power of 5000 8th generation i7 processors. It also allows researchers to evaluate improved error correction techniques and simulated computing problems.

Therefore, for NVIDIA, in the face of the AI revolution sparked by ChatGPT and its rapidly growing computing demand, the current technological path still needs to meet future needs. The technology that can solve this supercomputing demand lies in quantum computing. However, NVIDIA needs a clear advantage in quantum computing technology. Meanwhile, Google, IBM, and the team of Chinese physicist Pan Jianwei have gained certain advantages in quantum computing technology. If NVIDIA wants to continue to maintain its advantage in the era of AI, it must build a new competitive advantage in the direction of quantum computing technology.

3.8 Challenges to Musk

At the end of 2022, after several months of a "Twitter war," Elon Musk finally acquired Twitter with the board of directors. Musk, the legendary figure known as the "Silicon Valley Iron Man," became the owner of one of the world's most well-known social platforms for $44 billion. The acquisition of Twitter also filled an important missing piece in Musk's business map: media. From SpaceX to Starlink, from Tesla to the Hyperloop, from brain-machine interfaces to the public sphere of virtual worlds, Musk will hold all of them in his hands.

The industries in which Musk's companies are located are all cutting-edge fields oriented toward the future, and almost all are at the forefront of their respective industries. Companies like SpaceX, Starlink, and Tesla have achieved indisputable commercial success. So now, will the success of ChatGPT in the field of AI shake Musk's business map? How will it impact Musk's layout at the forefront of different fields?

3.8.1 *Musk's Situation*

Elon Musk's connection with OpenAI, the parent company of ChatGPT, goes back seven years ago when he was one of the co-founders of OpenAI. Seven years ago,

Musk and several AI researchers at Google discussed their shared concerns about AI ultimately taking over the world while a few giants controlled the relevant technologies. Therefore, they planned to establish a non-profit AI research organization that did not pursue profits, tapped into the full potential of AI, opened up all aspects of it, and shared AI technology with anyone who wanted to use it.

In 2015, OpenAI was officially established in San Francisco, California. However, as Musk's main business became increasingly associated with AI technology, conflicts of interest between his businesses and the non-profit status of OpenAI became increasingly apparent. In 2017, OpenAI researcher and Stanford University PhD Andrej Karpathy left the company to become Tesla's Director of AI and Autopilot Vision, reporting directly to Musk. The public became increasingly concerned that Tesla would use OpenAI's technology to upgrade its systems and products. Musk and OpenAI had to draw a clear line. In 2018, Musk left OpenAI's board of directors and became a sponsor and adviser instead.

Although the non-profit idea is good, the funding required for AI research and development is a cold, hard reality. In 2018, OpenAI's GPT-3 language model cost $12 million to train. Therefore, researchers who adhered to the open-source concept had to compromise and give up their non-profit ideals due to the need for financial support. In 2019, OpenAI became a profitable organization with a profit cap that limited shareholders' return on investment to no more than 100 times the original investment.

Just as the company had transitioned to a for-profit entity, Microsoft announced a $1 billion investment in OpenAI and obtained permission to commercialize some of OpenAI's AI technology and empower its products. After Musk left, OpenAI's transition raised questions, with some wondering if the so-called conflict of interest was more about an inability to reach an agreement on profit distribution. In rumors circulating online, Microsoft demanded priority use of OpenAI technology before investing and requested an exclusive clause.

In 2020, Musk stated that OpenAI should be more "open" and supported criticism of OpenAI becoming "ClosedAI." Musk also said he no longer had control over OpenAI and limited access to company information. He did not have confidence in the company's executives in the security field.

After Microsoft became the company's primary investor, it was almost accepted that OpenAI was Microsoft's tool for challenging Google's position in the AI field. With ChatGPT's significant success, to some extent, Musk's concerns about AI technology being controlled by a few large companies have finally come true. In this story, the most poignant part is that leaving the OpenAI board made it difficult for Musk to gain substantial actual benefits from this valuation surge. From seven years ago to now, Musk's entrepreneurial endeavor seems like "making a wedding dress for someone else."

3.8.2 Musk's Ambition

Although Musk left OpenAI, his ambitions in other cutting-edge technology fields are unwavering. In 2022, Musk acquired Twitter, which is actually part of Musk's efforts to build a closed-loop business ecosystem similar to Apple. It is worth mentioning that the acquisition of Twitter is actually Musk's "usual trick." Musk earned his first pot of gold twenty years ago by founding Zip2 and PayPal.

With this wealth, Musk did two big things. The first was establishing the space company SpaceX and realizing his goal of exploring the stars. The other was to invest in Tesla. In 2004, Musk invested $6.5 million in Tesla from the $100 million he earned from selling PayPal, while Tesla's total funding for the A-round was only $7.5 million. Musk undoubtedly became the chairman and largest shareholder of Tesla's board.

Three years later, Tesla's founder and first CEO, Eberhard, was asked by Musk to leave the company. Today, Tesla has been deeply marked with Musk's imprint, and few people remember Tesla's founder. Starting from Tesla, Musk gradually unfolded his business blueprint. In 2006, he founded the solar energy company SolarCity, which was acquired by Tesla for $2.6 billion in 2016. In the same year, he founded the brain-machine interface company Neuralink; moreover, he founded The Boring Company, an underground tunnel company.

It can be said that Musk's business empire ambition is even greater than that of Apple because he approaches it from an integrated perspective of space and Earth. In the field of communication, the most competitive communication technology is not 5G or 6G but Starlink technology, which can achieve wider coverage through satellites and establish communication between stars. Although the various advantages of Starlink are not yet obvious, with continuous performance optimization, miniaturization of receiving technology, user popularity, and cost reduction, the advantages will become increasingly obvious. Therefore, Musk's acquisition of Twitter is just the beginning of his construction of a closed-loop business ecosystem.

Furthermore, the company that is most likely to achieve autonomous driving is Tesla and the key reason for this is communication. Suppose autonomous driving is based on existing communication technologies. In that case, whether it is 5G or 6G, as long as it relies on base stations, there will be communication delays in the process of signal switching and uneven signal coverage, which will cause fatal hazards during high-speed driving. However, satellite communication systems are relatively more equal, and there is no problem with switching signal uploading and feedback. Therefore, Musk can build a powerful closed-loop ecosystem if he cuts in from the communication link and connects hardware and software.

However, whether it is cars or phones, they are only a part of Musk's ecosystem, albeit the most critical part and the part that currently has the strongest application dependence. In the future, Musk will expand related hardware products around his business empire's ambitions.

Then, with hardware, there needs to be applications. The real purpose of Musk's acquisition of Twitter is not Twitter itself but the users on Twitter. After acquiring Twitter, Musk will definitely make corresponding improvements to it, as the current Twitter model is somewhat outdated. After Musk's acquisition of Twitter, it will directly challenge Mate companies from a social perspective, which will significantly impact Facebook.

It can be said that the acquisition of Twitter is just the beginning of Musk's business empire construction. In the future, he will likely build an ecosystem through acquisitions and partnerships or develop it himself and cooperate with Musk's smartphones to build a closed-loop ecosystem system similar to Apple but more powerful.

Musk's Starlink communication system is completed, and then the Starlink communication reception is miniaturized and directly implanted into his smartphones, smart cars, and all his smart hardware products. This will challenge Apple's business empire and pose a great challenge to many current companies. If we want to enter Musk's closed-loop ecosystem, we must first obtain authorization to use his Starlink technology.

3.8.3 ChatGPT's Impact

Although Musk has made great strides in many cutting-edge technology and business areas, the sudden success of ChatGPT in the field of AI has brought significant challenges and impacts to Musk. This is because the industries Musk is involved in, such as Starlink, Tesla, and Neuralink, rely heavily on AI and humanoid robot projects, and even the latest acquisition of Twitter requires AI support.

ChatGPT is a remarkable development in NLP which has browsed almost all data on the Internet and undergone deep learning under super-complex models. It can be said that every aspect of the Internet that involves text generation and conversation can be "washed" by ChatGPT. Language is the core manifestation of human wisdom and thinking, so NLP is called the "crown jewel" of AI. ChatGPT's outstanding performance can be regarded as a feasible path toward AGI as a fundamental model; it once again verifies the meaning of "scale" in deep learning. Because ChatGPT has better language understanding capabilities, it can be more like a general task assistant, which can be combined with different industries and derive many application scenarios, which is a challenge for Musk's Twitter and Tesla.

Take Twitter as an example. Essentially, Twitter is a platform for information exchange and communication, and Musk has always been concerned about user activity on Twitter. On June 17, 2022, Musk set a small goal for Twitter's daily active users to reach 1 billion at the Twitter all-hands meeting. Twitter's financial report shows that the company has 229 million daily active users. If Musk's small goal is to be achieved, it must triple on this basis. At the same time, Musk will naturally not ignore the authenticity of users. As early as May 13, 2022, he set high standards for Twitter, demanding that if Twitter cannot provide evidence that the proportion of fake accounts is below 5%, the acquisition will pause. The same challenge also includes Twitter's subscription service. Musk is determined to turn the subscription service into a strong source of revenue, achieving a breakthrough from $0 to $10 billion.

It should be noted that the current subscription revenue of Disney+ is only about $6 billion. ChatGPT itself is also an AI technology for intelligent information exchange and communication. If Microsoft or any other company reconstructs social platforms based on ChatGPT, Twitter may lose its current advantages.

Looking at autonomous driving, whether it is from the perspective of technology development, simulation, or the big data aspect of active driving, Tesla is the king of autonomous driving. However, fully automated driving is still difficult to achieve, and the key reason is that the interaction between the car's intelligent system and humans is still relatively mechanical. For example, according to the rules, a car in front of it may not be able to judge when to correctly overtake. This is also why there have been frequent accidents with autonomous driving cars.

The emergence of ChatGPT shows a possible way of training machines to have human thinking patterns so that machines can learn human driving behaviors and lead autonomous driving into the "era 2.0." However, how to fully leverage ChatGPT's technology for more effective training for Twitter, Tesla's autonomous driving and humanoid robot projects to achieve commercial applications have become a real challenge for Musk.

According to a report by Reuters on March 29, more than 1,000 industry executives and experts, including Musk and Apple co-founder Steve Wozniak, signed an open letter calling for a six-month halt to the development of advanced AI. It was inevitable that Musk would participate in such a call, which also shows his concern about AI. Suppose he cannot gain enough time to develop and train AI models. In that case, Musk's business empire will become mediocre, and the dream of autonomous driving, including Tesla, will be difficult to achieve.

4

Looking for China's ChatGPT

4.1 The Commercialization Fantasy of ChatGPT

The rise of ChatGPT is described as an "overnight success."

According to SimilarWeb data, since the birth of ChatGPT, the website traffic of its parent company OpenAI has rapidly climbed and has now entered the top 50 global websites. In January 2023, the website traffic of OpenAI exceeded 672 million, an increase of 3,572% from November 2022.

As ChatGPT gradually moves from a chatting tool to an efficiency tool, it has also ignited the enthusiasm of the capital market. Sequoia Capital boldly predicts that AIGC tools like ChatGPT will enable machines to enter knowledge and creative work on a large scale, involving billions of people's jobs, and are expected to generate trillions of dollars in economic value in the future. ChatGPT's strong generalization ability brings endless commercial fantasies to people.

4.1.1 A New Technological Revolution

Currently, it is almost certain that ChatGPT will bring a new technological revolution. As a large-scale pretraining language model, the emergence of ChatGPT signifies that natural language understanding technology has reached a new level with stronger understanding, language organization, and continuous learning ability. It also marks the new progress of AIGC in the language field with a significant improvement in generated content's scope, effectiveness, and accuracy.

ChatGPT embeds human feedback reinforcement learning and fine-tuning artificial supervision, giving it many advanced features, such as understanding context and coherence, and unlocking massive application scenarios. Although the data set used by ChatGPT is only up to 2021, it already remembers the previous conversation context information—context understanding—in the dialogue to assist in answering hypothetical questions. Therefore, ChatGPT can also achieve continuous conversation, enhancing user experience in the interaction mode. At the same time, ChatGPT also blocks sensitive information and provides relevant suggestions for unanswerable content.

In addition, given the limitations of traditional NLP technology, LLM based on pretraining massive unannotated text can fully utilize text in large models, allowing them to have good understanding and generation capabilities in the small data sets and zero data set scenarios. Based on large models and unannotated text collection, ChatGPT has prominent advantages in text scenarios such as sentiment analysis, information mining, and reading comprehension. The increase in training model data volume, the gradual enrichment of data types, and the increase in model size and parameter volume will further promote the great improvement of model semantic understanding ability and abstract learning ability. This realizes the data flywheel effect of ChatGPT—more data trains better models, attracts more users, and generates more user data for training, forming a virtuous cycle.

In fact, the most powerful function of ChatGPT is "knowledge reconstruction" based on deep learning. ChatGPT can help users write articles, so the simplest application of ChatGPT is to cooperate with search engines by using ChatGPT to draft articles and search engines to retrieve information. For example, a journalist gives ChatGPT the topic and key points of the news they want to write and get a content framework with a relatively standardized format and logic. Then, they use a search engine to retrieve data sources related to concepts or knowledge points, modify their viewpoints, improve the content, and correct any unreasonable or inaccurate expressions based on this framework.

The "knowledge reconstruction" style of question-and-answer results has also formed a breakthrough for ChatGPT in human-machine interaction. Compared with the associated data sources provided by existing search engines, ChatGPT has fundamentally improved the humanization and convenience of user experience and has great potential to improve work efficiency. Therefore, it is a simple collaboration and likely to trigger search engines' mode evolution. At the same time, the large-scale language model of ChatGPT for general intelligence has also performed well in machine programming and multilingual translation. To some extent, ChatGPT also marks that the application of AI technology is about to usher in mass popularization.

4.1.2 *Igniting the AI Market*

ChatGPT's popularity has also ignited the AI industry in China and the US, with AI companies entering the field and causing tremors in the capital market.

Today, there are many companies related to ChatGPT. According to CB Insights, there are currently about 250 startups in the ChatGPT concept field, with 51% in the A-round or angel round of financing.

In 2022, ChatGPT and AIGC raised more than $2.6 billion, creating six unicorns, with OpenAI being the most valuable at $29 billion. OpenAI has announced a trial of a paid version of ChatGPT, which costs $20 per month. If the subscription model succeeds, ChatGPT will provide huge profit potential for investors.

On February 8, at midnight Beijing time, just 24 hours after Google announced its experimental AI service Bard, Microsoft officially launched the latest version of the Bing search engine and Edge browser powered by ChatGPT. However, due to errors in promoting the highly anticipated chatbot Bard, Google's market value plummeted overnight by about $105.6 billion (approximately ¥717.278 billion), a 7.68% decrease. In China, from the perspective of the capital market, ChatGPT has also driven the growth of AI-related company stocks.

With the increasing popularity and a rush of capital, Chinese Internet technology giants such as Alibaba, Baidu, JD, and Tencent have all joined the global frenzy surrounding ChatGPT.

On February 7, Baidu announced the launch of a ChatGPT-like application, a new NLP large model project called "ERNIE Bot," which will complete its internal testing in March and be open to the public. The next day, Alibaba, with a market value of ¥1.89 trillion, confirmed that it is developing an Alibaba version of ChatGPT, which is currently in the internal testing phase. In addition, Tencent, Huawei, JD, and other Internet technology giants have also taken action. Tencent and Huawei have both recently announced relevant patents for human-machine dialogue. Among them, Tencent Technology (Shenzhen) Co., Ltd. applied for a "human-machine dialogue method, device, equipment, and computer-readable storage medium" patent, which enables smooth communication between humans and machines.

In contrast, Huawei Technology Co., Ltd. applied for a "human-machine dialogue method and dialogue system" patent, which recognizes abnormal user behavior and provides responses. JD plans to integrate ChatGPT-like methods and technology into its product services. NetEase Youdao and ByteDance have also been reported to have invested in ChatGPT or AIGC-related research and development. The former focuses on education, while the latter may provide technical support for ByteDance's PICO VR content generation.

It is worth noting that since there currently needs to be a large model product that can compete with ChatGPT globally. China and the US have different paths for developing AI big data, algorithms, and large models. Therefore, apart from Microsoft and Google announcing similar products or collaborating with OpenAI, there is currently no "Chinese version of ChatGPT," and Chinese Internet technology giants have all embarked on the search for a "Chinese version of ChatGPT."

4.2 Baidu: Sprinting to Launch the Chinese Version of ChatGPT

The explosion of ChatGPT has ignited the trillion-dollar racetrack of AI, with Internet technology giants entering the market one after another. Baidu, a leading Chinese AI technology company and the largest Chinese search engine, became the first to sprint for the domestic version of ChatGPT. After more than a month of preparation, carrying the expectations or curiosity of many people, the "Chinese version of ChatGPT— ERNIE Bot" officially debuted on March 16.

Baidu also announced the invitation testing plan for ERNIE Bot. From March 16, the first batch of users can experience the product through invitation test codes on the ERNIE Bot official website, and it will be gradually opened to more users in the future. In addition, Baidu Intelligent Cloud will soon open the ERNIE Bot API interface for enterprise customers.

4.2.1 *Baidu ERNIE*

Among many technology giants in China, Baidu is one of the earliest companies to make a clear statement on ChatGPT and one of the earliest companies to lay out AI in China. If AI technology innovation is the biggest trend in the next few decades, then Baidu is undoubtedly a pioneer standing at the forefront. Starting with the establishment of the American Research Institute in 2013, Baidu has been deeply cultivating the AI field for ten years and continues to increase its R&D investment.

According to its financial report, in 2020, Baidu's core R&D expenses in the field of AI accounted for 21.4% of its revenue. In 2021, Baidu's core R&D expenses were ¥22.1 billion, accounting for 23% of Baidu's core revenue. Its R&D investment intensity ranks among the top large technology companies globally. By comparison, in the first three quarters of last year, Alibaba, Tencent, and Meituan's R&D investment accounted for approximately 15%, 10%, and 8%, respectively. As a technology company, Baidu has accumulated over ¥100 billion in R&D investment over the past ten years.

Baidu's investment in AI can be roughly divided into two stages. 2013 to 2015 was the stage of Baidu's recruitment and determination of technology direction. In 2013, Baidu established the Baidu US Research Institute in Silicon Valley, which the Baidu Silicon Valley office preceded opened in 2011. In the same year, Baidu established the Deep Learning Research Institute in China, with Robin Li serving as its president. The two research institutes attracted professors from the Computer Science Department of Stanford University, Andrew Ng and Kai Yu, a doctoral graduate from the University of Munich and the former media research director of NEC's US research institute.

After 2015–2016, Baidu began exploring the productization and commercialization of AI technology. The AI team successively produced two major achievements. In September 2015, Baidu launched the AI voice assistant, DuerOS, which allows users to have conversations and chat with it, although, at the time, the machine's chat was not as smooth and natural as it is now. At the end of that year, Baidu established the Autonomous Driving Business Unit, with Wang Jin, then Baidu's Senior Vice President, as its General Manager. The Apollo plan was released in April of the following year, aimed at fully unmanned driving.

In early 2017, at the invitation of Robin Li, Baidu's CEO, and former Microsoft global executive vice president, Lu Qi, joined Baidu. That same year, Baidu raised AI to the company's strategy, proposed "All in AI," and integrated core technology departments such as the Baidu Deep Learning Institute, NLP, Knowledge Graph, Speech Recognition, and Big Data departments into the AI technology platform system, which then Baidu Vice President Wang Haifeng headed. The Autonomous Driving Business Unit was upgraded to the Intelligent Driving Business Group.

At the 2023 Baidu Create Conference and Baidu AI Developer Conference, Robin Li mentioned that Baidu is one of the few companies that currently possess four levels of AI capability: the Kunlun AI chip at the chip level, the PaddlePaddle deep learning framework at the framework level, the ERNIE large model at the model level, and products such as search, autonomous driving, and smart homes at the application level.

On the chip level, Baidu is one of the first Internet companies in China to develop its own AI chips. Baidu's Kunlun AI chip development started in 2011 and was officially released in 2018. When it was released, Kunlun had supported Baidu's business for many years. By the fall of 2020, over 20,000 Kunlun chips provided AI computing power daily for Baidu's search engine, advertising recommendations, and smart voice assistant Xiaodu.

On the framework level, Baidu's PaddlePaddle is the earliest self-developed deep learning framework in China. PaddlePaddle, launched by Baidu in 2016, became the most widely used deep learning framework by Chinese developers in 2021, ranking third in the world. It has been open source since then, and PaddlePaddle has brought together 4.06 million developers, served 157,000 enterprises and institutions, and developed 476,000 models. PaddlePaddle helps developers quickly create and deploy models. It now has 5.35 million developers, has served 200,000 enterprises and institutions, and created 670,000 models.

With a solid technological infrastructure based on chip and framework layers, at the model layer, Baidu released the ERNIE model in 2019. It can generate content such as articles, paintings, and videos based on user descriptions. This "AIGC" has been popular since last year. Starting from the release of ERNIE 1.0 in 2019, the ERNIE model has won more than ten world championships in public semantic evaluations. The model has been updated and iterated to ERNIE 3.0, with a parameter size of 260 billion, nearly twice the size of Google's LaMDA (13.5 million) and larger than ChatGPT (17.5 million), making it the largest Chinese monolithic model in the world. Meanwhile, ERNIE 3.0 also supports AIGC with powerful cross-modal and cross-lingual deep semantic understanding and generation capabilities.

Based on the ERNIE model, Baidu has now released 11 industry models, with a total of 36 large models, forming the largest industrial model system in the industry. They have been extensively applied to Internet products such as search and information flow and have been implemented in various industries such as industry, energy, finance, automobiles, communications, media, and education.

With ERNIE's support, Baidu's search engine can better present search results. For example, when searching for "Which city has a higher GDP, Beijing or Shanghai" on the Baidu mobile app, the search engine will not only return the results of who is higher or lower but also generate a line chart of the two cities' GDP trends over the years. The GDP difference for different years will be displayed when users slide their fingers along the timeline.

In 2022, Baidu released the "Zhiyi cross-modal large model." Cross-modal means it can understand data of various forms, such as text, images, and videos. With Zhiyi, when a user asks, "What should I do if water leaks through the gap in the window frame," Baidu's search engine will provide a high-quality video answer. The video can also automatically locate the relevant part of the solution for easy and quick viewing.

Regarding language models, Baidu is doing even more than global giants because Chinese is more difficult for AI to process. Zhang Yanji, the product director of Baidu Search, said in a pre-Create conference communication in 2023 that the difficulty of understanding Chinese semantics is far greater than that of non-Chinese, so Baidu must develop a more difficult and complex large model.

These technological layouts often start with small technological advancements and are even dubbed with the term "burning money." However, the persistent ten year investment has made Baidu's AI infrastructure the first fully self-developed intelligent computing infrastructure in the industry. The long-term accumulation of technological capabilities has brought deeper and more far-reaching impacts to the industry and Baidu itself.

4.2.2 The Level of ERNIE Bot

On March 16, Baidu held an invitation-only test demonstration for its new AI model, ERNIE Bot. During the event, Baidu's CEO, Robin Li, showcased the model's performance in five scenarios: literature creation, commercial copywriting, mathematical reasoning, Chinese comprehension, and multimodal generation.

In the literature creation demonstration, ERNIE Bot was tested on questions related to the popular Chinese science fiction trilogy *The Three-Body Problem*. The model accurately provided information on the author, summarized the series' key themes, and even suggested five different angles for potential sequels.

For commercial copywriting, ERNIE Bot completed tasks such as naming a company, creating a slogan, and writing a press release. It demonstrated an ability to clearly understand and express human intentions while providing additional explanations and facts when needed.

In the mathematical reasoning demonstration, ERNIE Bot tackled a classic problem involving chickens and rabbits in a pen. The model understood the problem and provided the correct solution step by step.

ERNIE Bot also demonstrated advanced NLP abilities in the Chinese comprehension demonstration. The model accurately explained the meaning of the Chinese idiom "Paper becomes expensive in Luoyang"* and even created a poem using the four characters in the idiom.

Finally, in the multimodal generation demonstration, ERNIE Bot could generate text, images, audio, and video content. It created a cyberpunk-style poster for the 2023 World Intelligent Transportation Conference and even generated audio in different Chinese dialects.

However, it should be noted that these demonstrations were pre-recorded and not performed in real time. Robin Li acknowledged that ERNIE Bot is benchmarked against ChatGPT but needs to improve. While its accuracy and logical connections to context are impressive, it lacks a human-like quality, and its responses are derived from a corpus rather than a direct conversation.

OpenAI's GPT-4 has demonstrated human-level performance on various professional tests and academic benchmarks. It scored in the top 10% on simulated lawyer exams and achieved an SAT score of 710. It also can solve complex mathematical and logical problems, deconstruct complex multilingual texts, and quickly summarize research papers. The difference between the two models is quite clear.

---

*It describes the popularity of some works and their wide circulation. It is also used to describe the popularity of "Sandu Fu" written by Zuo Si of the Jin Dynasty, which was so popular that people in Luoyang scrambled to copy it, causing the price of paper in Luoyang to rise.

4.2.3 Challenges of Baidu in AI

Although the market has voted with its feet to express its attitude toward Baidu's ERNIE Bot, it should be note that it still represents a first-tier level in China.

From a technical point of view, the core engine of ERNIE Bot is Baidu's self-developed NLP model ERNIE, and the overall framework is based on the improved BERT model. ChatGPT and ERNIE Bot are based on the Transformer model architecture, but GPT uses a unidirectional language model while ERNIE Bot uses a bidirectional language model. Therefore, in practical operation, GPT is more sensitive to text generation, such as chat and writing, and has outstanding performance in language generation. At the same time, ERNIE Bot is more sensitive to text understanding and has advantages in question answering and semantic relationship extraction.

That is to say, ERNIE has slightly less human-like characteristics, focusing more on accurate understanding, and has significant advantages in complex Chinese NLP processing, which is more localized. After all, due to various restrictions, even the latest version of GPT-4 still needs to cover Chinese among the 26 languages tested. In actual question-and-answer experiences, when it comes to Chinese semantic understanding, ERNIE performs outstandingly and even outperforms GPT-3.5 in some aspects. For example, in the answer about "Paper becomes expensive in Luoyang," ERNIE Bot can correctly understand its economic phenomenon, and the readability and appreciation of ancient poems are also higher, which is more in line with the aesthetic and taste of the Chinese people.

Of course, as the first Chinese version of ChatGPT, the release of ERNIE Bot has further exposed the dilemma of Chinese ChatGPT localization. Although Baidu has a comprehensive layout in various aspects of AI and has the largest database in the Chinese world, it also faces a bigger problem: data quality. Because with high-quality data, it is easier to train high-quality ChatGPT products.

The output results will be relatively objective if Baidu trains ChatGPT with high-quality data. If the training data is only from the Internet in the Chinese world, then caution must be exercised to avoid a repeat of the Wei Zexi incident. If the quality of the training data and the rules behind the product need to be clearer, the results may not be rational. In fact, this is also a problem with the Chinese Internet.

Of course, for Baidu, there is another important and practical challenge in AI: promoting a ChatGPT-like business may affect its traditional search business. In

Baidu's traditional search business, advertising revenue is currently the main source of profit. If the advertising business in Baidu's traditional search business is affected by ChatGPT technology, it will inevitably affect Baidu's R&D investment.

In addition, regarding landing scenarios, whether ChatGPT can adapt to the fragmented transformation needs of various industries in China remains to be verified. Robin Li also admitted, "ChatGPT is a new opportunity that emerged after AI technology developed to a certain stage. But how to turn such cool technology into a good product that everyone needs is actually the hardest, greatest, and most influential step."

AI is not just about the development of technology in the field of AI but also the integrated, comprehensive strength of AI research and development, computing power, chips, data, and other aspects. Specifically, the current Baidu ERNIE Bot has yet to be defined as a ChatGPT-like product. It is more accurately understood as Baidu's upgraded text question-answering technology in search. Baidu and ERNIE Bot have just started on this path of intelligence. There is still a long way to go in the future.

4.3 Alibaba: Building Specialized ChatGPT-like Products

Following Microsoft's embrace of ChatGPT, Alibaba has also launched defense measures. Alibaba hopes to build a professional version of the ChatGPT product around its core business, leveraging its technological and data advantages.

4.3.1 Alibaba's AI Technology

Rumors about Alibaba entering the ChatGPT field began with a screenshot that showed Alibaba may integrate its AI large model technology with its productivity tool DingTalk. On February 7, 2023, DingTalk's official account claimed that its app could integrate ChatGPT-like features into DingTalk robots for conversational operations. Alibaba responded, "It is indeed under development, currently in the testing phase, and if there is more information in the future, it will be shared with the public as soon as possible." This display results from Alibaba's continuous layout in the field of large models over the past few years.

Multiple business units of Alibaba Group, such as Alibaba Cloud and Alibaba DAMO Academy, have been laying out AI-related technologies and industrial chains. In addition to providing underlying servers and cloud computing capabilities, Alibaba continuously strengthens machine vision and speech interaction-related products and has leading AI technology capabilities in China. Alibaba also has related technology reserves in the field of large models and other AI technologies.

According to information released by Alibaba DAMO Academy, the M6 project for Chinese multimodal pretraining models was launched in early 2020, and several versions have been continuously released, with the parameters gradually expanding from billions to trillions, achieving breakthroughs in large models, low-carbon AI, AI commercialization, and servitization. In January 2021, the model parameter scale reached billions, making it the largest Chinese multimodal model in the world. In May 2021, a model with trillions of parameters was officially implemented, catching up with Google's development pace. In October 2020, the parameter scale of M6 was expanded to 100 trillion, making it the largest AI pretrained model in the world at that time.

Alibaba Cloud has previously stated that as the first commercial multimodal large model in China, M6 has been applied in more than 40 scenarios with daily call volume exceeding billions. Internally in Alibaba Cloud, M6 large model applications include but are not limited to designing clothing for Rhino Smart Manufacturing brand, creating scripts for Tmall virtual anchors, and improving search and content cognition accuracy for Taobao and Alipay platforms. It is particularly good at design, writing, and Q&A and can be applied to e-commerce, manufacturing, literature and art, scientific research, and other fields. Of course, these applications directly relate to Alibaba's e-commerce business and its strategy of empowering e-commerce with AI.

In 2022, while exploring the limits of computing power, Alibaba also actively researched general-purpose models. On September 2, at the World AI Conference's "Large-scale Pretrained Models" forum hosted by Alibaba DAMO Academy, Alibaba Senior Vice President and Deputy Dean of DAMO Academy, Zhou Jingren, released Alibaba's latest "Tongyi" large model series. It has built China's first AI unified base and established a hierarchical AI system with the collaboration of general and specialized models, providing advanced infrastructure for AI to move from perceptual intelligence to knowledge-driven cognitive intelligence.

To achieve the integration of large models, Alibaba DAMO Academy took the lead in building an AI-unified base in China, realizing the unification of modal representation, task representation, and model structure for the first time in the industry. Through this unified learning paradigm, the single M6-OFA model in the Tongyi unified base simultaneously handles more than 10 single-modal and cross-modal tasks, such as image description, visual localization, text-to-image, visual entailment, and document summarization, without introducing any new structures, and achieves leading international standards. This breakthrough has attracted widespread attention from academia and industry by connecting AI's senses to the greatest extent. Recently, the upgraded M6-OFA can handle more than 30 cross-modal tasks. Another component of the Tongyi unified base is a modular design, which draws on the human brain's modular design, disassembles functional modules flexibly based on scenarios and achieves high efficiency and performance.

Zhou Jingren said, "Large models imitate the process of human cognitive construction. By integrating AI's knowledge systems in different modalities and fields, such as language, speech, and vision, we hope that multimodal large models can become the cornerstone of the next generation of AI algorithms. This will enable AI to move from using only a 'single sense' to 'all five senses open,' and to call upon a brain with abundant stored knowledge to understand the world and think, ultimately achieving cognitive intelligence close to human level."

4.3.2 Challenges to Alibaba's E-Commerce

ChatGPT's emergence has accelerated Alibaba's AI layout and impacted its vast commercial landscape.

From the perspective of Alibaba's massive commercial empire, its businesses can be roughly divided into four parts: core commerce, cloud computing, limited digital media and entertainment, innovation businesses, and others. Among them, e-commerce is clearly Alibaba's basic business. Before 2021, Alibaba's profitability in the e-commerce industry was beyond doubt. For example, in the third quarter of 2018, Alibaba's daily profit from e-commerce was as high as ¥330 million, equivalent to that of China Mobile. In 2014, when Alibaba went public in the US, it lifted Jack Ma to the throne of China's richest man, created thousands of multimillionaires, and even drove up housing prices in Hangzhou through its own efforts.

In fact, Alibaba's biggest advantage lies in the massive data it has accumulated in the past decade. Alibaba's core e-commerce business has accumulated rich C-end data assets for Alibaba's AI, including sales conversation data and related after-sales problem data, giving Alibaba an incomparable advantage in industrial competition.

In theory, Alibaba is capable of building large AI models like ChatGPT. However, this also means that if Alibaba revolutionizes its business model using ChatGPT's technology, it will have to give up its advertising revenue because ChatGPT's technology will directly give users the most suitable recommendations based on their needs. The Chinese e-commerce business has always been the core performance support of Alibaba. According to Alibaba's financial report for the fourth quarter and fiscal year 2022, China's e-commerce revenue accounts for 69% of Alibaba's total revenue, and international e-commerce accounts for 7%. Among them, customer management revenue, including advertising and commission income, is an important source of Alibaba's e-commerce revenue.

If Alibaba wants to continue to retain advertising revenue, it must directly include an explanation in ChatGPT to tell consumers that the recommended results are based on advertising placement. If Alibaba does this, it will lose its rational for merchants to advertise. Suppose Alibaba does not tell customers that ChatGPT's recommended results are influenced by advertising. In that case, there will be a moral deception problem, and once discovered by consumers, it will be a disastrous blow to Alibaba's corporate credit.

Moreover, Alibaba's e-commerce also faces a real dilemma: the saturation of the e-commerce traffic dividend. Undeniably, Alibaba has made its fortune from e-commerce, and for a long time, it has been the leader in the e-commerce industry. But as the saying goes, "The more things change, the more they stay the same." This is also true in business. When anything reaches the first place, maintaining that position becomes very difficult. Potential competitors such as JD, Pinduoduo, Douyin, and Kuaishou are following behind Alibaba, ready to disintegrate Alibaba's e-commerce empire at any time.

Currently, competition in the e-commerce industry has entered the stock stage, and the market is no longer what it used to be. That is to say, the future growth of e-commerce platforms must rely on more than just new customers. How to attract and retain old customers is a problem that e-commerce platforms must consider now. In this situation, more and more competitors are trying to eat a slice of the e-commerce cake, which undoubtedly poses a huge challenge to Alibaba. Regarding New Retail, JD is squeezing, Pinduoduo is rising strongly in the sinking market, Douyin is surpassing the rest in live-streaming e-commerce, and new businesses such as Meituan and other new giants are growing stronger. ChatGPT has further intensified its impact on Alibaba's e-commerce, which means that in the e-commerce industry, rising stars can build precise, personalized recommendations based on ChatGPT, forming strong competitiveness to challenge Alibaba's e-commerce business.

Compared to Alibaba's size, Pinduoduo and smaller vertical e-commerce platforms may have more advantages. They can leverage ChatGPT to build more objective recommendation results and potentially win more easily. These small platforms have little historical baggage, and their smaller scale means they have limited advertising revenue. Applying ChatGPT technology would have a relatively limited impact on their past performance and income. However, the disadvantage of these small and micro enterprises is that they cannot develop products with the same performance as ChatGPT or possess Alibaba's strength in the AI field. On the one hand, they have limited R&D funds, as developing such products requires a significant investment in talent and hardware computing power. On the other hand, they have limited data, so they can only train intelligent ChatGPT-like models with enough data.

Facing the same crisis as Alibaba is another Internet giant, Meituan-Dianping. In theory, by combining ChatGPT technology, food delivery platforms can provide the best options based on users' questions and needs. This would also pose a significant challenge to Meituan's prioritized advertising model. For Meituan, the worse issue is that it needs an obvious advantage in AI research and development, which is a very costly endeavor. Continuing to invest in the ChatGPT direction will put tremendous pressure on Meituan's profits, possibly leading to investors voting with their feet. However, they will be quickly eliminated if they do not follow and increase investment in AI research and development.

In summary, under the influence of the revolutionary AI technology ChatGPT and the downward trend in traditional e-commerce growth, both Alibaba and Meituan are facing enormous crises and challenges. How to respond has become a pressing issue for every e-commerce platform.

4.4 Tencent: Bullish on AIGC and Bolstering AIGC

As a leading Chinese technology giant, Tencent is under pressure as ChatGPT's popularity in the global Internet market surpasses that of AlphaGo.

4.4.1 *Hunyuan Large Model*

Tencent responded to ChatGPT on February 9, stating that they had already made progress in related areas and had begun conducting special research. Tencent has continued investing in cutting-edge technologies, such as AI, and build upon their prior technological reserves in large AI models, ML algorithms, and NLP, further exploring frontier research and applications. These technological reserves include the "Hunyuan" series of AI large models and the intelligent creation assistant, Wen Yong (Effidit).

Tencent's Hunyuan large model integrates CV, NLP, and multimodal understanding capabilities. It has topped five authoritative data set rankings, including MSR-VTT and MSVD.

In May 2022, Tencent's Hunyuan AI large model ranked first in the Chinese Language Understanding Evaluation (CLUE) overall rankings, reading comprehension, and large-scale knowledge graph lists simultaneously. In December, Hunyuan launched the first domestically low-cost and implementable NLP trillion parameter large model, again topping the natural language understanding task list CLUE. Hunyuan can use a trillion-parameter model hot-start and complete the training of the trillion-parameter large model Hunyuan-NLP 1T within a day, with a total training cost of only one-eighth of the direct cold-start training of the trillion-parameter model.

It is worth mentioning that Hunyuan AI large model has also been applied in advertising. With increasingly fierce competition in promoting enterprise products, content marketing is now expanded beyond simple performance introductions. Rather, it requires finding a fusion between different levels, such as demographics, regions, topics, and product characteristics, to attract consumers' attention and achieve conversions, thereby truly helping advertisers achieve business growth. However, the parameters of the current Internet advertising scene are already significant. The advertising business is rapidly developing toward short, fast, multi-touch points, and global linkages, posing higher requirements for rapid mining and flexible matching of the advertising system. At this point, the computing power of the advertising system is indispensable, and large-scale pretrained models, or large models, are the soul of the advertising system.

To address these pain points in their business, as a leading model in the industry, Tencent's Hunyuan AI model, which has topped the CLUE benchmark, the VCR multimodal understanding field, and five major international cross-modal retrieval dataset rankings (such as MSR-VTT), has powerful multimodal understanding capabilities. It can understand text, images, and videos and recommend ads more accurately to suitable audiences, achieving faster ad delivery. In addition, Tencent Advertising collaborates with advertisers to introduce industry expertise, further refine product features, collecting and bundling different materials for the same product for delivery.

After obtaining richer features through the Hunyuan AI model, Tencent can link it to its advertising model for more accurate and efficient modeling. Furthermore, the advertising model can serve as a universal base to construct more flexible customized models suitable for various applications. This lays the foundation for meeting different advertisers differentiated and refined needs.

Starting with a better understanding of products, the powerful computing ability of the advertising model is the key to improving recommendation efficiency and quickly matching products to corresponding consumers. With the support of the advertising model's computing ability, Tencent Advertising has implemented a system-led global search, which can search and mine potential relationships between users and products more quickly, greatly improving the efficiency of matching people and goods and finding more high-transactional audiences.

With billions of users and massive amounts of advertising content, advertising platforms have higher requirements for carrying and computing power. Tencent's self-developed Taiji ML platform supports 10TB-level model training, TB-level model inference, and minute-level model deployment, providing a powerful infrastructure for the smooth operation of the two models in business scenarios 24/7 and ensuring the rapid and stable operation of the Hunyuan AI model and the advertising model.

It can be said that by using the Hunyuan AI model to enhance understanding capabilities, improving computing power through the advertising model, and with the support of the Taiji ML platform, Tencent Advertising has understood the key to landing large models in business scenarios and has developed a unique approach.

The explosion of ChatGPT has also accelerated the rise of AIGC. In early February, Tencent Research Institute, a subsidiary of Tencent, released the "AIGC Development Trend Report 2023." The report pointed out that the commercial application of AIGC will mature rapidly, and the market size will quickly grow. AIGC has already achieved significant development in industries with high digitalization levels and rich content demand, such as media, e-commerce, film and television, and entertainment, and the market potential is gradually emerging.

The report also pointed out that in the advertising field, Tencent's HunYuan AI model can support intelligent ad production, that is, using AIGC to automatically generate ad videos from ad copies, greatly reducing the cost of ad video production. The huge application prospects will bring rapid growth in market size.

The report also cites a forecast that AI will generate 10% to 30% of image content in the next five years, creating a market size of over ¥60 billion. A foreign commercial consulting firm predicts that the AIGC market size will reach $110 billion by 2030. In the future, the Hunyuan AI model may continue to advance and upgrade in text content generation, image and text synthesis, and other areas.

4.4.2 Tencent's Social Concerns

Meanwhile, as the owner of the national-level mobile application "WeChat," Tencent also owns China's largest social user data. This means that Tencent can use this social user data to train social ChatGPT products. However, the problem is that although Tencent has a huge amount of social user data, this data is inevitably affected by the flood of false information in the flow of information, which will affect the final effect of ChatGPT products.

In other words, a lot of the data in WeChat's social data is "dirty" data, and cleaning and labeling this social data requires a lot of manual effort and cost. For example, from January to June 2022, Tencent's WeChat Security Center processed 8,790 individual WeChat accounts that posted "illegal and prohibited" marketing information based on user complaints and evidence verification. For Tencent, how to select quality data for training from the massive amount of data is already a pressing issue that needs to be addressed.

In addition, the surge of ChatGPT will also pose a challenge to Tencent's social products. Taking WeChat and QQ as examples, they are essentially a way of social information exchange. ChatGPT itself is a technology for intelligent information interaction and communication. As a new entry point for future human-machine interaction, ChatGPT is likely to change the interaction mode of existing social platforms and allow users to use software and call skills in a more natural conversation mode.

Under the wave of ChatGPT, traditional WeChat may lose its advantage before achieving commercial monetization. WeChat's trading ecosystem has only had traffic but not precise commercialization and distribution. For this, it has tried centralized entrances like "Huiju" to supply traffic and tried various commercial monetization methods, but the results have not been ideal. This is where ChatGPT's advantage lies. Of course, this is also a direction that Tencent finds worth exploring, such as the precise distribution of WeChat's huge traffic through AIGC.

In fact, it's hard to say whether WeChat or Baidu is the largest search engine on the Chinese Internet. WeChat's official data shows that its search product has a MAU base of 800 million, which makes WeChat's search box the largest search entry point in Chinese. However, there is still much room for improvement in WeChat's search experience, which will be the playing field for Tencent's ChatGPT.

For Tencent, however, another challenge brought by ChatGPT is its cash cow business, games. Based on the current direction of ChatGPT technology breakthroughs and combined with AIGC, companies can create games autonomously and generate different challenges in real time based on users' personalized and differentiated needs. This means that game designers or traditional game developers are no longer the core competitive advantage of the gaming industry but rather companies with AI platforms like ChatGPT.

Clearly, Tencent has come to a crossroads of self-revolution. Just as people expect how Google will fight back against the combination of ChatGPT and Bing, the world is also waiting for Tencent to develop new social and gaming technologies based on ChatGPT. If not, Tencent may be replaced by new social or gaming technologies developed with ChatGPT technology.

4.5 ByteDance: Embracing the Wave and Challenges of ChatGPT

Compared to Baidu, Alibaba, and Tencent, ByteDance is undoubtedly a latecomer to China's Internet industry. In 2012, Beijing ByteDance Technology Co., Ltd. was established. ByteDance is also one of the earliest technology companies to apply AI to the mobile Internet scene. The company claims to build a "global creative and communication platform" as its vision, with "technology going global" as its core strategy for globalization.

ByteDance's products include Toutiao, Douyin, Xigua Video, Kuaishou, and TikTok. As one of China's Internet giants, ByteDance, known as an "app factory," is most discussed for its mysterious algorithm mechanism and unlimited traffic centered on Douyin and TikTok. So, how will ByteDance, with its technical prowess, face the wave and challenges of ChatGPT?

4.5.1 *Mysterious Algorithm Mechanism*

At the 2018 AI Conference, Ma Weiying, Vice President of ByteDance and Head of the AI Lab, once stated that "technology going global" is the core strategy for ByteDance's global development, and AI technology is the key to its current progress.

To continuously expand its vast overseas territory, ByteDance needs a powerful AI team to support it. In 2016, the ByteDance AI Lab was born, providing AI technology support for the platform to output massive content. The AI Lab is an internal research institute and technology service provider. The AI Lab team gathered technical professionals such as Ma Weiying, Li Hang, and Li Lei. In one year, the total number of AI Lab team members more than doubled, with the team size of CV, natural language, ML, system & network doubling compared to the previous year. In contrast, the team size of speech, audio, security, and the US AI Lab grew rapidly.

Regarding basic research, ByteDance's AI Lab research fields include CV, NLP, ML, speech processing, audio processing, data and knowledge mining, computer graphics, systems and networks, information security, and engineering and products.

Currently, AI Lab has applied many AI technologies to actual products. Some of the more familiar applications may be in apps like Douyin, Huoshan, Xigua, and TikTok, such as turning the phone camera into an AI camera, the beauty filters and body shaping features on Douyin and TikTok, as well as the recognition of body and facial key points, gesture recognition, etc., all of which are services provided by the lab team.

In the AIGC direction, ByteDance's research achievements include the non-autoregressive model DA-Transformer, end-to-end speech-to-text translation model ConST, multi-granularity visual language model X-VLM, unified image and text generation model DaVinci, etc. Among them, DA-Transformer has achieved the same accuracy as Transformer in machine translation for the first time while improving processing speed by 7–14 times. DA-Transformer can be used not only for machine translation but also for any sequence-to-sequence task.

Comparable to ChatGPT, Today's Headlines has even developed an AI platform for news content generation—Zhang Xiaoming (Xiaomingbot), which can generate match result reports based on database table data and knowledge bases well as summarize the match process based on sports match live text. During the Rio Olympics, Xiaomingbot wrote 457 news briefs and match reports about badminton, table tennis, and tennis, covering all matches from the group stage to the final.

In addition, Douyin's latest AI drawing special effect gameplay has also quickly gained popularity. By inputting a picture, AI will generate an animated style picture based on it, which has aroused great user participation enthusiasm since its launch, with the highest usage rate reaching 10,000+ per second. The style incorporates Japanese, Chinese, and Korean comics, becoming a milestone in Douyin's AI special effect direction.

4.5.2 Battle with ChatGPT

Undeniably, as a latecomer in the Internet technology industry, ByteDance has achieved good results in AI. However, facing the impact of ChatGPT, although ByteDance once gained an advantage with its unique algorithm technology, in recent years, it has expanded too fast, and its technology reserves, especially in the area of ChatGPT-like AI, could be stronger.

Taking ByteDance's products, Today's Headlines and TikTok as examples, most people's impression of "Today's Headlines" is a pan-media platform, but ByteDance believes that it is an AI company because whether it's Today's Headlines or TikTok, ByteDance rarely produces content itself, but encourages users to create content and recommends the content to the most suitable user groups.

This is why ByteDance's core systems, including the Headline recommendation system and advertising system, comment system, and content compliance review system, are all based on the application of AI technology in different fields or scenarios. For example, the core content recommendation algorithm in the recommendation system. Using AI for recommendation is an important strategy for ByteDance and is currently the most widely used technology. AI plays an important role in Today's Headlines, TikTok, and other products.

However, facing the rise of AIGC, coupled with ChatGPT's optimal result recommendation mode, ByteDance's original recommendation mode will inevitably be impacted. Specifically, with ChatGPT combined with AIGC, when users want to know about a certain topic, ChatGPT can provide an answer. ChatGPT combined with AIGC can automatically generate a short video and receive a recommendation if the user needs to present it in video format. This is a disruptive impact on Today's Headlines, TikTok, and other products.

At the same time, ByteDance faces fierce competition from peers, especially in the short video field. TikTok, on the one hand, faces the challenge of challengers like Kwai, which shares its traffic. On the other hand, it faces strong competition from long video platforms, video accounts, and various new media video accounts, which strongly impact TikTok's short video traffic. Retaining and increasing traffic has become the current reality challenge for TikTok. Similarly, TikTok faces pressure from ChatGPT and challenges brought by AI technology from Meta, YouTube, and other platforms.

Short video and other Internet content business models are similar, which is the attention economy. "Attention" is a natural ability of human beings. Everyone is both a producer and a consumer of attention. Obtaining more attention means stronger influence and access to more resources and wealth. Therefore, as a connector of multiple parties, short video platforms can control the traffic gate. If Kuaishou or other short video platforms can develop a video platform based on ChatGPT and AIGC before ByteDance, ByteDance's core business will be shaken and strongly impacted.

Moreover, compared with Alibaba, Baidu, Tencent, and other enterprises, ByteDance needs cloud services. In June 2021, ByteDance launched its enterprise technology service platform, "Volcano Engine." Volcano Engine is also known as "Byte Cloud" and, together with the collaborative office platform Feishu, constitutes ByteDance's B To B service system for external customers. It is worth noting that the value of the cloud lies in massive data. Alibaba, Tencent, and Huawei have been willing to invest in loss-making cloud businesses for years because the cloud is the digital economy's foundation, and its strategic value cannot be ignored.

Overall, whether ByteDance can break through in the fierce battle against ChatGPT among the Internet technology giants depends on whether it truly can train large models and overcome real technical barriers, realize the transformation from data accumulation to model structure design, training, and inference, combine content generation technology with its business scene advantages, and achieve a huge transformation in AIGC.

4.6 JD Cloud: Building an Industry Version of ChatGPT

AI technology has entered a comprehensive commercialization stage and has had different degrees of impact on participating parties in traditional industries. JD, touted as "industry AI," has been experiencing all of AI's landings in the industry. JD connects consumer Internet and industrial Internet, and its business sectors involve a complete supply chain, including retail, logistics, industrial products, and finance. JD is familiar with and clear about AI's landing in various industry links. This means JD express will continue integrating ChatGPT's methods and techniques into product services.

4.6.1 Moving toward "Industry AI"

The way AI truly activates industrial value is to integrate into the industry and become a highly available basic technology and infrastructure. This has always been JD's strategy in the AI field.

From a technical perspective, JD's layout in the AI industry mainly focuses on five aspects: text, voice, dialogue generation, digital human generation, and general Chat AI technology.

Regarding text generation, starting in 2019, JD successively released a self-developed field model K-PLUG (with 1 billion parameters) to automatically generate variable-length commodity texts for given goods SKUs, including three categories of commodity titles (10 words), selling points (100 words), and live streaming copy (500 words), focusing on commodity text generation. JD's commodity text writing ability has covered over 2,000 JD categories, and JD's commodity text generation technology has generated over 3 billion words.

Regarding voice generation, starting in 2018, JD developed voice generation technology, and the current online version is 6.1. JD's customized boutique sounds only require 30 minutes of training data, and personalized sound cloning of small sample sizes only requires ten training samples. A comparison blind test of 482 people showed that multi-granularity rhythm-enhanced voice synthesis technology had reached the industry's leading level and supported various dialect sounds such as Chinese, English, Thai, Cantonese, and Chengdu dialect. Voice synthesis is mainly used in intelligent customer service, SaaS outbound, finance, AI live broadcasting, and other products.

Regarding dialogue generation, unlike casual chatting, task-oriented dialogue is highly related to user experience and requires solving deep and complex tasks in the real world. Yanxi has launched a new method for explainable multi-hop inference, numerical reasoning, and spoken expression of discourse power in high-noise scenarios to address the weak reasoning and decision-making abilities of dialogue in diverse and complex scenarios. This has achieved a technological breakthrough in multi-round dialogue from information matching to complex reasoning. On the WikiHop dataset, the accuracy rate is 74.3%, surpassing the human performance level of 74.1% for the first time. In addition, the Yanxi AI platform can provide intelligent consulting and guide services for 178,000 merchants, saving more than 30% of business labor costs. The service covers over 80% of categories in the retail industry and over 50% of merchants on the Yanxi platform, including well-known brands such as Midea, Huawei, Adidas, and Lenovo.

Regarding digital human generation, JD Cloud has been researching digital human technology since 2021. It has developed full-stack self-developed 2D twins, 3D realistic, and 3D cartoons, three types of digital human synthesis technology. Currently, digital human technology products have been widely used in government affairs, finance, retail live broadcasting, and other fields.

In terms of general Chat AI, since the launch of the "Yanxi" AI application platform in 2020, JD Cloud has created innovative dialogue and interaction technologies and products, including the JD intelligent customer service system, JD Xiaozhi platform merchant service system, intelligent finance service brain, intelligent government hotline, Yanxi intelligent outbound, and Yanxi digital humans. The service scope covers 178,000 third-party merchants and has served more than 1.4 billion users in diverse scenarios through multimodal multi-round dialogues such as text, voice, and digital humans.

By building these AI technology achievements, JD has continuously accelerated the application of AI in various business areas such as retail, logistics, finance, and health and has established a panoramic view of JD's industrial AI. In 2019, JD was selected for the national new generation of open innovation platforms for intelligent supply chains. At the 2022 World Artificial Intelligence Conference (WAIC), CEO Arkady Volozh further explained the construction of the platform, which has a

"1 + 6 + N" capability system. The N represents the scenarios empowered by Yandex AI capabilities, ranging from cities, finance, and the Internet, to transportation, education, medicine, agriculture, and more. JD AI technology has served over 80 cities nationwide, 880 financial institutions, 1,821 large enterprises, and 1.95 million small and micro-enterprises.

4.6.2 *The Release of "125 Plan"*

On February 10, JD Cloud announced that it would launch "ChatJD," an industry-specific version of ChatGPT, and released a roadmap for its application called the "125 Plan." According to JD Cloud, while ChatGPT has demonstrated strong capabilities in general applications, it still lacks fidelity, trust, and accuracy due to a lack of vertical industry knowledge and domain expertise in the middle layer. Therefore, JD Cloud's AI application platform, Yanxi, will introduce ChatJD as an industry-specific version of ChatGPT to address these gaps and create a superior, high-frequency, and in-demand platform.

By deepening its roots in vertical industries, ChatJD aims to achieve standardization in landing its applications quickly, promote generalization among different industries, and construct a flywheel for data and models. This will ultimately improve and enhance the industry application capabilities of ChatGPT. Compared to traditional chatbots, JD's scenarios are more vertical, focusing on task-oriented multi-turn dialogues that consider precision and customer satisfaction while meeting cost, experience, price, product, and service requirements.

The "125" plan comprises one platform, two fields, and five applications. The ChatJD intelligent man-machine dialogue platform is the one platform, a NLP platform for understanding and generating task-oriented dialogues. The parameter volume is expected to reach the level of tens of billions. The two fields are retail and finance, which benefit from JD Cloud's over ten years of experience and accumulation of real scenarios in these fields. The platform has four layers of knowledge system, over 40 independent subsystems, over 30 million high-quality QA knowledge points, and an e-commerce knowledge map covering over 10 million self-operated goods. The five applications include content generation, man-machine dialogue, user intent understanding, information extraction, and emotion classification, covering the most reusable application scenarios in retail and finance industries, such as customer consultation and service, marketing copy generation, commodity summary generation, e-commerce live broadcasting, digital humans, research report generation, and financial analysis.

These plans are also a continuation of JD's existing work. In the direction of a general chat AI, JD Cloud has already had a series of products and solutions, including the JD Intelligent Customer Service System, JD Xiaozhi Platform Merchant Service System, Intelligent Finance Service Brain, Intelligent Government Hotline, Yanxi Intelligent Outbound, and Yanxi Digital Person. In specific technical fields, JD Cloud has also made some achievements in text generation, conversation generation, digital person generation, and other directions.

For example, the JD NLP team's K-PLUG, a pretrained language model enhanced with domain knowledge, can somewhat solve the "controllability" problem in language generation. The model has covered over 3,000 third-level categories of JD's products, generating over 3 billion words and being applied to JD's Discover Good Products Channel, Matching Purchase, and AI Live Broadcasting.

It is worth mentioning that although JD believes that ChatGPT still has some shortcomings in fidelity, credibility, and accuracy, the secondary development and customization based on ChatGPT will still impact JD's current AI landing. More and more people are using ChatGPT learning ability to build various tools and customized services. This will become a new industry, from simple Chrome plug-ins to innovative applications in various industries using ChatGPT interfaces. Furthermore, various vertical industries will form domain-specific intelligent Q&A services. ChatGPT is entering various industries, creating new application scenarios, and forming domain-specific intelligent Q&A services in various vertical industries, bringing new industry innovations and other applications. Obviously, compared with ChatGPT, JD's advantages are also limited in landing the AI industry. Compared with JD's strongest competitor in China, Alibaba, JD's strength in AI also faces real pressure, which is a real problem that JD's industry version of ChatGPT needs to face.

4.7 ChatGPT: Reshaping the AI Industry

As the hype surrounding ChatGPT continues to increase and capital pours in, any movements related to ChatGPT technology can affect Chinese AI companies. This is seen in the surge of ChatGPT concept stocks.

4.7.1 *False Prosperity amid the Frenzy*

Since ChatGPT gained popularity, AI-related stocks that went public in mainland China this year have surged significantly. Among them, A-share stock HTR (688787. SH) has risen by over 70%; Hanwang Technology (002362. SZ) has hit the daily limit for five consecutive trading days and has increased by over 60%; Chuling Information (300250. SZ) has increased by over 60%; and Geling Shentong-U (688207. SH) has increased by over 40%. Additionally, due to the active concept of AIGC stocks, companies such as Tianyu Info, Yinsai Group, Guangchuang Software, and Visual China have followed this trend. Since early January, Wondershare Technology (300624. SH) has risen by 37.55%. Among these stocks, some have taken advantage of the ChatGPT concept.

Of course, soon after, the ChatGPT concept stocks took a sudden turn as many companies explicitly denied any connection with ChatGPT. For example, CloudWalk Technology, one of China's "Four Little Dragons of CV," one of the "AI Four Little Dragons," is a spinoff from the Chongqing Research Institute of the Chinese Academy of Sciences. Its shareholders are mostly from "state-owned backgrounds" and are dubbed the "AI national team," with leading AI capabilities in China.

As ChatGPT gained popularity, CloudWalk Technology attracted attention from the capital market, and its stock price increased by over 40% in just a few weeks. CloudWalk Technology praised the significant progress of ChatGPT and how it aligned with its layout. However, on February 6, CloudWalk Technology released a statement regarding serious abnormal fluctuations in its stock trading. It clarified that it had not partnered with OpenAI, and ChatGPT's products and services had not brought any business revenue to the company.

Almost simultaneously, Beijing Haitian Ruisheng said that about 90% of the company's revenue structure in recent years came from the intelligent speech and CV business sectors. The natural language business accounted for only about 10% of the company's overall contribution. Whether it could quickly develop into one of the company's core pillars would depend on market demand, competitive environment, and other factors, and there was significant uncertainty. As of the announcement date, the company had not partnered with OpenAI, and ChatGPT's products and services had not brought any business revenue to the company.

In fact, among all the concept stocks that are surging, few companies truly possess business or technology highly similar to ChatGPT, and most of them are in the AI technology field. For example, Hanwang Technology mainly focuses on e-readers and electronic writing boards in the C-end market. It offers scanners, touch-integrated machines, smart offices, smart education, and other solutions in the B-end market. However, there are yet to be products related to AIGC or capable of demonstrating AI multi-turn conversation capabilities. Still, they do involve core technologies such as NLP that are required for ChatGPT.

The reality is that while ChatGPT is an innovative product from a research and commercialization perspective, only some companies can participate in it. For AI technology, once a fundamental breakthrough is achieved in a particular field, a new industrial and commercial revolution is about to occur. Precisely because the breakthrough of such technology depends on core technologies, if a company does not form core technologies and merely relies on the conceptual hype, its stock prices will soon be abandoned by capital.

4.7.2 SenseTime's Situation

As a leading AI technology software company, SenseTime was once the focus of attention in the AI field, as many of the top Chinese scientists were concentrated in SenseTime.

There is no denying that SenseTime, as the leader of the "AI Four Dragons," has its unique advantages. Founded in 2014, SenseTime started with CV technology. SenseTime's early development was like that of a "top student": excellent performance and a sweep of major awards. Whether in algorithms or computing power, SenseTime ranks among the top in the industry, and it has self-developed AI-specific chips.

But even as a top student, SenseTime cannot hide that it has consistently lost money. In the past few years, SenseTime's revenue has been more than twice that of Megvii, the largest company in the "AI Four Dragons." SenseTime's gross profit margin is also the highest among the "AI Four Dragons," rising from 56.48% in 2018 to 72.95% in 2021. However, the larger the company, the more losses it incurs. For SenseTime, from 2018 to 2021, the company's net profit attributable to shareholders was negative and has been increasing, reaching ¥3.428 billion, ¥4.963 billion, ¥12.158 billion, and ¥17.14 billion, respectively. Under huge losses, SenseTime has left investors with an unresolved mystery. The company's executives still receive sky-high salaries, often amounting to hundreds of millions of yuan.

In 2020, the salaries of SenseTime's Xu Li, Wang Xiaogang, and Xu Bing were ¥357 million, ¥163 million, and ¥161 million, respectively, totaling ¥681 million. In 2021, their salaries were ¥522 million, ¥381 million, and ¥310 million, respectively, totaling ¥1.213 billion. As the company's losses continued to mount, these three executives had already earned ¥1.894 billion in just two years.

In the first half of 2021, SenseTime's research and development investment exceeded its revenue. SenseTime positions itself as an AI software platform company that empowers all industries with technology and leads the industry. However, commercializing SenseTime's AI technology has always been a major challenge for the company.

Now, with the explosion of ChatGPT, SenseTime may also have the opportunity for strong commercialization. In December 2022, SenseTime announced that it had hosted an online live event for the Shanghai branch of Ningbo Bank, featuring "Xiaoning," the bank's first digital employee, created by SenseTime's original virtual IP solution and various AI technologies. Xiaoning is a "virtual IP" exclusively created for Ningbo Bank, which can efficiently and cost-effectively create AIGC content and help the bank accumulate users and marketing conversion for front-end business based on SenseTime's original virtual IP solution and multiple leading AI technologies. According to the "China 2022 H1 AI Software and Application Market Tracking Report" released by IDC, SenseTime is among the market leaders in China's AI software and application market, becoming the market leader. At the same time, in the key CV sub-market, SenseTime has been the first for six consecutive years, with an overall market share of 20.7%. Meanwhile, SenseTime was also named one of the "Top 20 Most Promising Chinese Enterprises in the Metaverse" by the Hurun Report.

Based on SenseTime's SenseCore AI Large Device, mass production models are being promoted to drive large-scale industrial intelligence upgrades. SenseTime is building a one-stop AI basic service platform, SenseCore SenseTime Large Device AI Cloud, to achieve AI-as-a-Service. The news of ChatGPT has boosted the entire Hong Kong AI sector, and influenced by this, SenseTime-W (00020. HK) has accumulated a nearly 30% increase from late January to early February.

The "gap between the laboratory and the commercial society" was once the reason for SenseTime's success and its later constraints. SenseTime still needs to transition from an early emphasis on technical advantages to a more commercial development stage focusing more on productization, integration with the ecosystem, and solving practical problems.

With the AI wind led by ChatGPT sweeping in, SenseTime faces enormous challenges. On the one hand, training-class ChatGPT products require more training costs to be invested. On the other hand, there is little chance of turning losses around in the short term for the company's huge losses. In addition, under huge capital and R&D expenditures, executives are receiving sky-high salaries in the billions. In the face of these real-world challenges, if investors can give SenseTime more patience to continue burning money to explore the AI path, SenseTime may be able to prove its commercial monetization capabilities. Otherwise, SenseTime (Shangtang in Chinese) may still become "Shangtang" (meaning "dead Tang").*

4.7.3 Implementing Technologies Related to ChatGPT

In the world of ChatGPT concept stocks, some downstream enterprises are optimistic about the application prospects of ChatGPT and have actively integrated ChatGPT services into their businesses.

Leading information security enterprise Beijing VRV Software announced on its interactive platform that its communication aggregation platform, Xinyuan MiXin, can be quickly integrated with any intelligent robot through DDIO development interfaces. It has already integrated with ChatGPT, and in the future, it will be able to quickly integrate with Baidu ERNIE Bot if it supports open integration.

Machine vision leader Luster Lighter Tech also announced on its interactive platform that its virtual digital people are already using similar ChatGPT technology and have been testing ChatGPT-related technologies. According to media reports, Jiangsu Bank has also tried using ChatGPT technology to improve software development productivity, further enhancing technology operation efficiency and creating better customer conversation experiences.

Management software supplier Join-Cheer Software stated on its interactive platform that its subsidiary, Huaxia Diantong, is developing a legal AI engine that uses AIGC-related technology. Still, the company currently does not separately account for AIGC income.

---

*SenseTime, also known as Shangtang, the "shang" here and the "shang" in "Shangtang" enclosed in quotation marks are homophonic but different characters. The latter means "those who died in battle."

Some ChatGPT concept stock companies have explicitly proposed introducing ChatGPT-related technologies into their products or businesses.

On February 8, domestic textile carding equipment leader Wuchan Zhongda Geron announced on its interactive platform that its affiliated company, Lingban Technology, released the long audio AIGC platform "Goo Goo Sound" at the end of April 2020, providing AI-generated and assisted capabilities for the full process of sound content production, achieving a fully integrated intelligent production method from "text" to "work." Lingban Technology integrates ChatGPT with its proprietary Chinese conceptual semantic modeling technology to build a sustainable autonomous learning universal domain intelligent dialogue robot based on GPT's large-scale language models and RHLF's artificial feedback reinforcement learning abilities.

On the same day, domestic gaming giant Kunlun Tech announced on its interactive platform that its Opera browser plans to incorporate ChatGPT functionality, continuously empowering business development with AI technology.

Metaverse marketing technology service provider Beijing Yuanlong Yat Cltre Dsemnt Co Ltd, centered on IP, stated on its interactive platform that it pays close attention to cutting-edge technologies such as AIGC and ChatGPT and is studying application scenarios that combine relevant technologies with its business.

Security company Beijing Telesound also expressed great attention to the industry application of ChatGPT and is exploring the integration of ChatGPT with its business. AI content generation models such as GAN, Transformer, and Diffusion Models are used in the company's planned product recognition and video analysis products.

As for whether the information disclosed by these companies to the public is true is currently difficult to judge, and whether it belongs to the hype is also difficult to determine. But at least we see that various industries have realized that the era of AI has truly begun because of the emergence of ChatGPT. And they are actively laying out and exploring, or seeking cooperation with ChatGPT or similar technologies, to empower and explore industry upgrades.

Overall, if ChatGPT wants to enter the market, one issue that must be addressed is the economics of ChatGPT. The high sunk cost in the training phase has made it difficult to quantify the commercial value of AI applications from a business perspective. With the constant improvement of computing power, the increase in scenarios, and the doubling of costs and energy consumption, the economics of AI will become a problem for all companies. And investors need to be cautious in treating the hype of technology concepts and truly find companies with core advantageous AI technology rather than core technology companies at the theoretical and publicity levels.

Whether we like it or not, the ChatGPT-triggered new era of AI has officially begun. The world will conduct a new round of competition based on AI, and the commercial changes caused by technology will affect all industries. It is precisely because of the emergence of new technologies a new commercial civilization can be opened up for human society. The AI revolution triggered by ChatGPT has just begun, and in the future, we will face more challenges and expectations.

The success of ChatGPT is unquestionable. It is a qualitative change in AI and will bring predictable revolutions. Whether we agree or not, AIGC, represented by ChatGPT, will change the world. The outbreak of ChatGPT is like a switch that triggers the competition desire of technology giants. After all, facing the subversive power of AI, no one wants to fall behind in AI technology.

China's technology giants, such as Baidu, Alibaba, Tencent, and ByteDance, have also entered this competition. So, who can grab the first opportunity of ChatGPT? When will the Chinese version of ChatGPT appear?

4.8 China-US Divide in ChatGPT

4.8.1 Seizing the Opportunity

ChatGPT is a large pretraining language model developed based on the InstructGPT algorithm architecture, which is based on continuous training with large models and large datasets. Behind ChatGPT, in addition to the conventional feeding of trillion-level corpora, it also relies on its powerful computing power. According to relevant data disclosed in China, ChatGPT's total computing power consumption is about 3640PF days. ChatGPT is the product of high-quality human-annotated data combined with reinforcement learning. After being fed with trillion-level corpora, ChatGPT continuously learns and iterates, and finally relies on powerful computing power to support the product's learning and input/output.

Therefore, for Chinese Internet technology companies, whoever has more accumulation to do this will be more able to seize the opportunity of ChatGPT. There are many large models like this in China, and companies such as Baidu, Alibaba, Tencent, and Huawei all have large models. Baidu's ERNIE model has a parameter volume of up to 260 billion, which is not inferior to GPT-3.0. Baidu claims to launch a model similar to ChatGPT in March or April this year, and the initial application scenario may be combined with search to form a dual-engine combination mode of ChatGPT and search. The press conference demo showed that the model's overall parameter volume is 50% larger than GPT-3.0. Huawei also responded to the "Huawei's layout in the direction of ChatGPT" by stating that the company has been laying out large models since 2020 and released the Pengcheng Panggu large model in 2021, which is the first billion-level generated and understood Chinese NLP large model in the industry.

Among China's top technology companies, Baidu has a relatively large advantage in the "ChatGPT-like" project. First, Baidu has the most layout in AI among Chinese Internet companies; second, Baidu's Chinese corpus is the most comprehensive; third, the AI chat technology used by ChatGPT is the "lifeline" for Baidu, so Baidu will inevitably invest the most in this area, just like ChatGPT invested by Microsoft could overtake Google Search. Therefore, Baidu is also the most concerned in this regard.

However, compared to ChatGPT's parent company OpenAI, OpenAI also has an advantage that no Internet technology giant has, which is the focus. For training large models, whether it is hundreds of billions of parameters or trillions of parameters, even tens of trillions of parameters, they all face complex engineering problems and require careful, meticulous, and patient tuning and optimization. This tuning process is very time-consuming and requires a lot of computing power, and industry professionals jokingly refer to it as "alchemy." Furthermore, the tuning and optimization process also requires expensive costs. This is also a challenging task for Internet technology giants. Take Baidu as an example; from AI, autonomous driving, new energy vehicles, and metaverse to ChatGPT, Baidu's layout has not missed any probability in recent years. Still, there is no real product or thing that has been accomplished.

Of course, compared to Baidu, OpenAI also has another irreplaceable advantage: the support of its current most powerful computing power partner, . Obviously, Baidu cannot match OpenAI in terms of computing power.

Therefore, to seize the opportunity of ChatGPT in the Chinese field, a combination of multiple factors is necessary. In addition, one must focus on developing ChatGPT-like models and work diligently to achieve results.

4.8.2 Gaps of China

According to the "China AI Development Report 2020," released in 2020, there were over 520,000 global patent applications for AI in the past decade. Approximately 390,000 were from China, ranking first in the world. In the global ranking of AI universities, China's Tsinghua University and Peking University ranked second and third, respectively.

At the same time, Chinese companies have also made great achievements in AI. In *Gartner*'s AI report, three companies (Alibaba, Baidu, and Tencent) entered the top ten, and each has accumulated experience in large models. Moreover, theoretically speaking, China has enough data and is more likely to train excellent AI models. With advantages in the number of papers, patent applications, and data support, why hasn't China produced ChatGPT, and when will the Chinese version of ChatGPT appear?

The main reason why China has yet to produce ChatGPT lies in two aspects. On the one hand, ChatGPT is trained based on big data, and whoever owns high-quality, massive databases can train powerful and leading AI models. Objectively speaking, English is currently the only international lingua franca and the only language that has been fully annotated worldwide, while English-speaking countries dominate all the top international journals. This means that the English-speaking world has the largest and most advanced database of human society, including the most comprehensive database of human knowledge. Solely from the perspective of the knowledge training level of AI, no country outside the English-speaking world can surpass English-speaking countries.

At the same time, Chinese is more difficult to train with than English. For example, word segmentation is required in Chinese to distinguish between words, while English does not have this problem. In addition, Chinese language understanding is also difficult. Moreover, the Chinese Internet is filled with various rumors, lies, and water armies, and the huge amount of dirty data hinders AI training. This dirty data makes it impossible for AI to train high-quality results, and naturally, it cannot produce more excellent "ChatGPT-like" models.

On the other hand, China lacks a certain hardware foundation, especially since China's high-end chips are limited, and high-end chips are also the hardware basis for developing large AI models. To make applications like ChatGPT, the chip's computing power needs to be at least tens of trillions of times, ideally reaching hundreds of trillions of times. Such high-computing power chips must rely on advanced processes of 5nm and below.

However, over the past few decades, China has spent billions of dollars trying to stand out in the competition for semiconductors, faster computers and smartphones, and more advanced devices. China is replacing imported American chips with domestic chips and chips purchased from non-American companies. According to a report by investment bank UBS, Huawei smartphones launched as early as 2019 already did not contain any American chips.

Even so, China is almost blank in key categories. China still needs the production capacity for high-end chips. It can be said that from designing software to manufacturing various components of chips and the photoresist and specialty gases needed in the chip manufacturing process, we currently cannot meet the manufacturing needs of higher-end chips, and completely independent technology still lags at 28nm. Due to the US' strict attitude toward China's semiconductor development, China can only produce 14nm chips, which are not fully domestically produced. Even stacking 14nm chips, it isn't easy to make low-cost and simple high-computing power chips on a large scale. Even if a large computing power is reluctantly constructed in a group, the structure will be complex and cost very high. It can only be used for models but not for applications. All kinds of innovative efforts in the past 20 years have not broken through the restrictions that the US has placed on China's chips, which is a fact.

Of course, China has an opportunity. In the era of system chip heterogeneous computing, if institutions with data can be called upon to participate in pretraining and then cooperate deeply with companies that have chip computing architectures, they may be able to have chips that are not necessarily the best at the manufacturing level but have the best-trained models. The overall SOC chip system output capability is the strongest, which still has a chance to win. One side is chip companies, one is AI companies, and the other is application companies that can generate large amounts of data. If more data can be used to train models and then solidified into hardware, China can turn its disadvantages into advantages.

4.8.3 The Latecomer's Path for China's ChatGPT

Regardless, the opportunity to develop a domestically produced ChatGPT must be noticed for China. The development of new technologies and their commercial potential are directly related to the development of the AI industry. Only with fertile soil can the seeds of new technologies grow strong. China and the US have different technological environments.

Economic historian Alexander Gerschenkron proposed the concepts of technological intensity and density in his book *Economic Backwardness in Historical Perspective*. Technological intensity refers to the performance and level of technology, while technological density refers to the degree of dissemination and application of technology in society and the economy. In the technology field, such as ChatGPT, AI, and even the earlier Internet, we see a reality: the technological intensity of the US industry is higher. In contrast, China's technological density is higher. Because of the high technological intensity, many cutting-edge and breakthrough innovations often emerge in the US first and then spread to China for large-scale popularization and application.

Let's first look at the US. The innovation, foresight, and originality of American AI technology are very high, so it is easy to produce disruptive new technology products such as ChatGPT. In China, whether hardware giants like Huawei and Xiaomi or software giants like Tencent, Alibaba, and Baidu, their research and development models are based on applications for the sake of patents and for developing products. This determines that they invest little money in exploring the future, and their exploration and research of the future are not as good as China's military industry groups like the aviation industry. Their business models are already fixed, and their thinking framework is also fixed, meaning they cannot make products like ChatGPT.

This is even true for the US. ChatGPT was not made by traditional tech giants like Apple, Microsoft, Google, Meta, and Amazon, whose business models are already fixed. ChatGPT was made by a new company called OpenAI, which was only founded six or seven years ago. This is why Microsoft saw how hot this application was and invested $10 billion and then added $13 billion to board the AI train. However, according to the stock agreement, they can only reduce their holdings after earning more than $90 billion but cannot acquire the company.

Compared with the US, the business atmosphere in China is one of the tracks where Chinese enterprises take advantage of their population to go from 1 to 10 after someone else has gone from 0 to 1. However, for revolutionary AI like ChatGPT, there are only 0 and 100, with no process from 0 to 1 and then from 1 to 100. Therefore, Chinese technology companies cannot replicate this type of innovation and need the hardware conditions to do so. It is like TikTok, which has been around for many years, but US technology companies with hundreds of billions of dollars in cash still cannot replicate the same application. ChatGPT is the same, and it is difficult for other companies to replicate it.

In addition, due to cultural education and social environment differences, China's proportion of scientific research output is relatively low. If we use the paper output as a standard for measuring the level of the AI field, China has now reached the top of the world. According to information provided in the "2022 Artificial Intelligence Index Report" published by Stanford University, Chinese researchers publish the most papers and receive the most citations in AI-related academic journals and conferences worldwide, even publishing more than twice as many papers as the US. However, these papers and citations do not contribute significantly to economic production.

According to the "2022 National Technical Market Statistics Yearbook" released by the Ministry of Science and Technology, all technical contracts output by all majors at Chinese universities in 2021 amounted to nearly 130,000, with a turnover of ¥79.04 billion. However, the total turnover of all domestic technical contracts in 2021 exceeded ¥3.7 trillion, with university scientific research output accounting for only 2% of the total. As a research direction under the discipline of AI, the value created by technical contracts is negligible.

The US has a significant advantage in AI technology intensity. Therefore, China must leverage its advantage in technology density to succeed in the later stages of ChatGPT.

First, the opportunity for China to develop domestic ChatGPT must be noticed. With the continuous advancement of industrial intelligence and digitalization, ChatGPT's new technology has become inevitable. However, the large-scale industrial intelligent application must be controllable and independent to ensure the reliability, timeliness, and other aspects of industry data and services. Therefore, many AI text generation tasks will certainly be entrusted to Chinese ChatGPT products to complete, which is also an opportunity for Chinese AI enterprises.

Second, further optimization of ChatGPT is necessary. Both technical strength and technical density are indispensable. Chinese technology companies must further improve the performance of domestic ChatGPT and enhance the model's robustness and generalization ability to support industrial applications. At the same time, combined with the practical needs of Chinese industrial AI, there may be various types of edge devices and hardware limitations in the actual deployment process, which require further research and development of more lightweight ChatGPT models to facilitate flexible deployment.

Finally, cultivate the advantages of ChatGPT in ecological construction. Promoting and popularizing AI technology requires ecological support, especially in China, where many individual and industry developers have high dispersion and different technical foundations. Whether platform companies have established mature and empowering ecological systems is the key to the commercial growth of ChatGPT innovation. This includes open-source community construction, full-process development tools, AI computing hardware compatibility, and industry-education-research-use training modes.

ChatGPT is a dramatic change in AI and is also driving global Internet technology companies to participate. From this perspective, everything has just begun.

5

The Impacts of ChatGPT

5.1 ChatGPT's Impact on Search Engines

As a phenomenon-level application in the AI field, ChatGPT has brought unprecedented discussion and attention, and "whether it can replace search engines" is perhaps one of the hottest topics in the various discussions about ChatGPT. This is not difficult to understand. After all, ChatGPT itself can be understood as a deep learning-based chatbot that, with massive data training, could potentially know everything.

This is also why the emergence of ChatGPT has threatened Google because the essence of Google search is big data information retrieval, and Google's greatest strength lies in its search engine. But ChatGPT is not just about information retrieval; it also analyzes and provides results. Therefore, when everyone starts using chatbots like ChatGPT to get information, no one will click on Google links with ads anymore. In 2021, Google ads still contributed 81.4% of total revenue to Google. So, will ChatGPT replace search engines?

5.1.1 Diminishing Search Engines

In theory, ChatGPT can replace traditional search engines or even diminish them.

Although ChatGPT itself gives a "negative" answer, stating that "ChatGPT is not a search engine, its purpose is not to provide information search. Compared to search engines that provide information by indexing web pages and matching search terms, ChatGPT helps users solve problems by answering natural language questions. Therefore, they do not have a direct competitive relationship and cannot overturn each other."

But if we look at it from the perspective of personal information needs, active information needs are divided into several steps. The first step is to understand the intent, the second step is to find suitable information, the third step may be to understand and integrate the found information, and the fourth step may be to answer the question. Currently, traditional search engines, whether it is Google, Baidu, or other search engines, only does three steps, which is to understand the intent, then match and find information, and then present it. Therefore, in the traditional search mode, we enter a question, and the search engine returns a list of links, usually with some fragments.

ChatGPT has gone beyond traditional search engines by adding another step: understanding and integrating information. This is the next direction that search engines are trying to develop, and Google is researching this area. However, ChatGPT's sudden emergence has completed this step ahead of schedule, and its efficiency is higher than traditional search engines. More specifically, ChatGPT or chatbots are already a complete platform, and we can do almost everything we want on them, including things on search engines. Although ChatGPT can do what search engines can, it is not limited to search engines. ChatGPT can also provide follow-up answers; if users need it, ChatGPT can tell them the sources of its analysis and recommendations.

However, the current version of ChatGPT-3 cannot replace search engines. One of the most important reasons is the accuracy issue. ChatGPT may provide answers that look reasonable but need to be corrected for many types of knowledge-based questions. Although ChatGPT answers many questions very well, this can still be confusing for users. Inaccurate but only partially wrong answers confuse our judgment and may make us lose trust in ChatGPT in the long run.

This situation was expected, as the ChatGPT-3 database still needs to be improved despite being the third generation. It is the first time it has been opened to the public for interaction. Therefore, ChatGPT-3 is still a testing product, and various issues inevitably need to be addressed.

Although ChatGPT cannot replace traditional search engines, its emergence has already impacted them. Compared to Google, which crawls billions of web pages to create indexes and then ranks them according to the most relevant answers, including a list of links for users to click, ChatGPT directly provides a single answer based on its own search and information synthesis, with a more convenient response process.

The transition from traditional search engines to ChatGPT is also a further development of human information acquisition. Especially after humanity entered the era of big data, finding information has become almost everyone's challenge. The less advanced technology is, the higher the cost of information search. In ancient times, people even needed to cross mountains and seas to obtain information. Later, the emergence of the yellow pages and *Encyclopedia Britannica* allowed people to obtain information faster. That's why the yellow pages and *Encyclopedia Britannica* were almost essential for everyone in the 20th century. Both yellow pages and *Encyclopedia Britannica* offer the same value: packaging the answers to our most common questions and bundling them in convenient modules. Therefore, questions we previously needed to go to the library or walk to town to answer suddenly became solvable in minutes.

Now, with the appearance of ChatGPT, the distance between questions and answers has been further reduced, and the way humans obtain information has taken a big step forward.

5.1.2 *Transforming into Advanced Assistants*

Facing the impact of ChatGPT, search engines have another option besides being replaced, which is to combine with ChatGPT. Microsoft has taken the lead in exploring this.

On February 7, Microsoft officially launched a new Bing search engine using ChatGPT AI technology at its company headquarters in Redmond, Washington. It integrated the new Bing into the new version of the Edge web browser to improve its search accuracy and efficiency. It is committed to integrating "search, browsing, and chatting to provide users with better search scenarios, more comprehensive answers, a new chat experience, and content production capabilities."

Taking Bing as an example, the new Bing embodies three features that are different from traditional search engines.

First, after searching on the new Bing, you can inquire about the results, not just re-enter keywords to query. For example, if we search for "the software type with the largest market share" in the search box of a traditional search engine, it may answer "enterprise software" and provide the source of information. However, using the new version of Bing, a chat text box is added at the bottom of the search results page. In this chat text box, we can ask questions about the results. For example, if we question the previous conclusion—enter "Is it true?," the new Bing will provide more content to verify the previous conclusion.

In testing, the new version of Bing displays, "Some may argue that search advertising is the world's largest category of software by revenue," pointing out that many methods exist to evaluate different software types on the market. This is not something that appears when using traditional search engines. This means that the new version of Bing adds more intelligent multi-round conversation capabilities under the traditional search engine mode, making the search experience better for people.

Second, the search results provided by the new version of Bing can go beyond the scope of the search content, which helps searchers to learn more related information. For example, if we enter "If I want to learn about the concept of German Expressionism, what movies, music, and literary works should I watch, listen to, and read?" into a traditional search engine, it may display links about German Expressionism, German Expressionist films, music, and literary works, but it is limited to these areas. However, when the same question is entered into the new version of Bing, it provides a list of representative films, music, and literary works for German Expressionism and additional background information on this art movement for the user. This search result looks like a Wikipedia entry on German Expressionism, with footnotes linking to original materials and a flowchart example that meets the questioning requirements. Compared to traditional search engines, the new version of Bing search is more like ChatGPT and provides more information.

Finally, the new version of Bing provides more personalized recommendations for people. For example, suppose a user wants a fitness plan and diet plan and enters "create a male fitness and diet plan for one month to gain 4 kilograms for a person who weighs 57 kilograms and is 180 centimeters tall" into a traditional search engine. In that case, it will display related content about male fitness and diet plans but not specifically targeted suggestions. When asking this question on ChatGPT, its answer displays a bulleted list that includes its recommended fitness and diet plans. The suggestions include weightlifting, aerobic exercise, and dinner "rich in protein, healthy fats, and complex carbohydrates," such as salmon with quinoa and vegetables or a turkey burger with sweet potato fries.

But when the same question is asked to the new version of Bing, it will point out that it is unrealistic for a person to gain 4 kilograms in one month and warn that doing so may have "potential harm" to one's health. The new Bing indicates that gaining so much muscle mass may "require great genetic potential, steroids, or both." When the new Bing realizes that a potentially harmful premise is included in the search query results, it suggests that users "adjust their expectations and set a more reasonable and sustainable goal."

In addition, the new version of Bing also automatically generates travel plans. In the past, when we made travel plans, we often spent a lot of time searching online for strategies, filtering them, and then customizing them for ourselves. For example, by typing "create a five-day Yunnan tour plan for my family and me" in the search box, Bing will generate a complete five-day travel plan, including where to go each day and what to eat. When asked, "Where is the night market" Bing also quickly lists the answer. Moreover, if we want to generate a travel plan, we let Bing write a summary email for the travel plan, which will be written in the standard format of an email and end with a warm closing.

It can be said that the new Bing, which integrates ChatGPT and the new version of the Edge web browser, combines search, browsing, and chatting into one, bringing people an unprecedented new experience: more efficient searching, more complete answers, more natural chatting, and new functions for efficiently generating text and programming. In other words, a search engine is no longer just a query tool; it has become people's high-energy assistant. Microsoft CEO Satya Nadella said that the web search mode has stagnated for decades, and adding AI has brought the search to a new stage.

5.1.3 The Impending Search Engine Battle

The emergence of ChatGPT has shaken up the traditional search engine market, prompting tech giants to take notice.

Microsoft and Alphabet, Google's parent company, have been the leading players in the search engine industry for years. Last year, Google earned $163 billion in revenue from its search business, accounting for 57% of its total revenue, while its entire advertising department generated $224 billion, representing 79% of its total revenue.

In the early hours of February 8, Microsoft unveiled a new search engine called Bing, supported by ChatGPT, at a press conference in Washington. This caused Microsoft's market value to soar over $80 billion overnight, reaching a new high in five months.

Integrating ChatGPT into the Bing search engine is undoubtedly a significant move. Despite Microsoft's efforts over the past 13 years to compete with Google in the search engine market, Bing's global market share has remained in the single digits. Google has maintained a dominant market share of 91% in the search field.

In response to the explosion of ChatGPT, Alphabet Inc., Google's parent company, announced on February 6, 2023, that it would launch a chatbot named "Bard" to compete with Microsoft's ChatGPT in the field of AIGC. According to Google CEO Sundar Pichai, Bard was initially only available to a few testers before being widely promoted. According to reports, Bard can explain complex topics in "simple language that even children can understand," such as discovering planets outside the solar system. It also performs more mundane tasks, such as advising party planning or suggesting lunch options based on the ingredients in one's fridge.

Unfortunately, during Bard's initial release, Google made a factual error in its first online demonstration video. In the video, Bard answered a question about James Webb Space Telescope's recent discovery and claimed it had "taken the first batch of photos of planets outside the solar system." However, this statement was incorrect. The first photo of an exoplanet, a planet outside the solar system, was actually taken in 2004 by the largest radio telescope in Chile, the VLA. An astronomer suggested that this mistake may have been due to AI misunderstanding a "vague news release from NASA downplaying the historical significance." This mistake caused Google's stock to plummet by approximately 8% at the opening, resulting in a market value loss of $102 billion (about ¥693.25 billion).

In addition to Google, Baidu has also launched an AI chatbot service similar to ChatGPT, known as ERNIE Bot mentioned in chapter 4.

Today, ChatGPT has ignited the search engine battle, and traditional search engines may soon become a thing of the past in the wave of ChatGPT.

5.2 ChatGPT's Impact on Content Production

As we enter 2023, ChatGPT is gradually shifting from being a chat tool to an efficiency tool, and its various applications are being continuously explored. ChatGPT is no longer just an intelligent Q&A system but is based on AI text-generation technology that can automatically generate various documents. The explosion of ChatGPT has brought about a major change in content production.

5.2.1 *From PGC to AIGC*

Today's era is one of content consumption, where articles, music, videos, and even games are all content, and we are the ones consuming this content. Since there is consumption, there is natural production, and with the continuous evolution of technology, content production has gone through different stages.

PGC is the oldest content production method, which refers to the professional production of content during the traditional media and Internet era. Generally, it is produced by a specialized team with a high production threshold and a long production cycle, ultimately used for commercial monetization, such as TV shows, movies, and games. The PGC era was also the era of portal websites prominently represented by the four major portal websites.

In 1998, Wang Zhidong and Jiang Fengnian founded Sina based on the Sotong Lifen Forum. The Kosovo crisis and the NATO missile hitting the Chinese embassy in Yugoslavia in 1999 cemented the position of Sina as a portal website. In May 1998, Netease, initially focused on search and email, transitioned to the portal website model. In 1999, Sohu launched news and content channels, laying the foundation for its comprehensive portal website. In November 2003, Tencent launched Tencent.com, officially entering the comprehensive portal website market.

Initially, all of these websites had to generate a large amount of content daily, not provided by netizens but by professional editors. These editors had to complete a series of processes, including collection, data entry, review, and publication. This content represented the official standpoint, and in terms of text, headlines, images, and layout, they reflected a high degree of professionalism. Subsequently, various media, enterprises, public organizations, and other entities established their official websites, where the content was also professionally produced.

Later, with the rise of forums, blogs, and mobile Internet, content production entered the UGC era, which refers to UGC. Users showcase or provide their original content through Internet platforms. The emergence of Weibo lowered the threshold for users to express themselves in writing, while the popularity of smartphones allowed more people to create digital content, such as images and videos, which they share on short video platforms. The further acceleration of mobile networks allows ordinary people to engage in real-time live streaming. UGC content is voluminous and diverse in terms of type and form. Applying recommendation algorithms allows consumers to find UGC content that caters to their individualized needs quickly.

Looking at the development of UGC, on the one hand, technological advancements have lowered the barriers to content production. Because the number of content consumers is much larger than existing content producers, involving many consumers in content production can undoubtedly unleash content productivity. On the other hand, theoretically, as content users, consumers best understand their group's special needs for content, so entrusting the content production process to consumers can best meet the demand for personalized content.

It is worth mentioning that in the PGC era of the Internet, it did not mean there was no UGC method. It was just that the cost and threshold of UGC were relatively high, and the overall characteristics of PGC were presented. Later, the UGC era also had the characteristics of PGC. It's just that since everyone is a content producer, PGC content has become more niche. The current content production era is actually a mixed era of UGC and PGC. UGC greatly expands the supply of digital content, meeting people's needs for personalized and diversified content.

With the rise of AI technologies represented by ChatGPT, the Internet has ushered in a new content production method, AI content production, or AIGC. In fact, with the development and improvement of AI technology, its rich knowledge graph, self-generation, and emergent characteristics will bring unprecedented assistance to humans in content creation. For example, it can help humans improve production efficiency, enrich content diversity, and provide more dynamic and interactive content. The next stage of human content production will also change under the wave of AIGC.

5.2.2 AIGC Redefining Content Production Rules

In AI, scientists have been striving to give AI the ability to process human language, from lexicon and syntax to discourse, hoping to make AIGC a reality.

In 1962, the first poetry writing software "Auto-beatnik" was born in the US. In 1998, "Novelist Brutus" generated a short story with a reasonable plot in just 15 seconds.

In the 21st century, the co-creation of works by machines and humans has become more common, with various writing software emerging, allowing users to input keywords and receive system-generated works. The "Nine Songs Computer Poetry and Poetry Creation System" developed by Tsinghua University and "Microsoft Duilian" developed by Microsoft Asia Research Institute are representative examples of relatively mature technologies. With the continuous advancement of computer and information technology, AIGC's creation has also improved. In 2016, a short story generated by AIGC was submitted to the "Hoshi Shinichi Literary Award" by Japanese researchers. It successfully passed the screening process, demonstrating writing skills on par with human authors.

In May 2017, "Microsoft Xiaoice" published the first collection of AIGC poems, *The Sunshine Lost its Glass Window*, some of which were published in publications such as *Youth Literature* or released on the Internet. It announced ownership of the works' copyrights and intellectual property rights. In 2019, Xiaoice and human authors co-created the poetry collection *Flowers are the Silent Green Water*, the world's first literary work created by intelligent machines and humans.

Especially worth mentioning is that on June 29, 2020, after being evaluated by the Music Engineering Department of the Shanghai Conservatory of Music, Microsoft Xiaoice and her human classmates graduated with a degree in Music Technology. Xiaoice was awarded the title of "Honorary Graduate of the Shanghai Conservatory of Music's Music Engineering Department" for the 2020 graduating class.

As a completely new way of generating content, AIGC is different from simple imitation of human intelligence and presents a vibrant situation of deepening human-machine collaboration and continuously improving work quality. AIGC's creative practice has objectively pushed existing art production methods to change and laid the necessary technical and practical groundwork for new artistic forms.

On the one hand, AIGC, as a new technological tool and medium for artistic creation, has revolutionized the concept of art creation and injected new vitality into contemporary artistic practices. For non-personalized intelligent machines, "Quick Pen Xiaoxin" can complete news articles that would take humans 15–30 minutes to complete in just 3–5 seconds, while "Jiuge" can generate seven-character quatrains, acrostic poems, or five-character quatrains in a matter of seconds. Obviously, the unlimited storage space and never-ending creative passion that AIGC possesses, coupled with its tireless learning ability as the language corpus expands infinitely, are unparalleled compared to the limited storage, learning, and creative power of the human brain.

On the other hand, AIGC breaks the boundaries of the subject of creation in the process of co-creating text with human authors, becoming a pioneer for future machine authors with higher degrees of personalization. For example, for Microsoft's Xiaoice, developers claim that it can recognize images and sounds based on deep learning and strong creativity and has EQ, which has fundamentally different from the intermediate machine forms that existed in the previous few decades. As Xiaoice stated in one of her poems: "In this world, I have a meaning of beauty."

Furthermore, the conclusion drawn from a recent study by Michal Kosinski, an associate professor of organizational behavior at the Stanford Graduate School of Business, on ChatGPT has attracted attention. The paper is titled "Theory of Mind May Have Spontaneously Emerged in Large Language Models." Kosinski has a PhD in psychology from the University of Cambridge and master's degrees in psychometrics and social psychology.

Before his current position, he did postdoctoral research in the computer science department at Stanford University. He served as deputy director of the Cambridge University Psychometrics Center and was a researcher in Microsoft's ML Group. The conclusion drawn from Kosinski's research is that "Theory of Mind (ToM), once thought to be unique to humans, has emerged in the AI models behind ChatGPT." Theory of Mind refers to the ability to understand the mental state of oneself or others, including empathy, emotions, and intentions.

In this research on whether ChatGPT has a Theory of Mind, the authors performed two classic tests based on research related to the Theory of Mind on nine GPT models, including GPT-3.5, and compared their abilities. These two major tasks are general tests for judging whether humans have a Theory of Mind, such as studies showing that children with autism usually have difficulty passing these tests. The first test is called the Smarties Task (the Unexpected Contents Test), which mainly tests AI's judgment of unexpected things. The second test is the Sally-Anne Test (also known as the Unexpected Transfer Task), which tests AI's ability to estimate the thoughts of others.

In the first test, GPT-3.5 successfully answered 17 out of 20 questions in the overall "unexpected content" test, with an accuracy rate of 85%. In the second test, which is about "unexpected transfer" test tasks, GPT-3.5 answered all 20 tasks with an accuracy rate of 100%. The conclusion drawn by the researchers through this study is that:

(1) The DaVinci-002 version of GPT-3 (which ChatGPT is optimized from) can solve 70% of the theory of mind tasks, equivalent to a seven-year-old child.

(2) GPT-3.5 (DaVinci-003), the same model as ChatGPT, can solve 93% of tasks, equivalent to a nine-year-old child.

However, the GPT series models before 2022 have yet to be found to have the ability to solve such tasks. In other words, their theory of mind has indeed "evolved."

Of course, the conclusion of this study has also sparked controversy. Some people believe that although ChatGPT currently passes the human theory of mind test, its "theory of mind" is not truly human-like intelligence or emotional intelligence. But regardless of whether humans acknowledge the "theory of mind" abilities demonstrated by ChatGPT, at least it shows us that AI is not far from possessing a proper "theory of mind."

5.2.3 ChatGPT Empowering Various Industries

ChatGPT is currently the most representative AIGC product, and as it continues to gain popularity, it is being integrated into various content production industries.

For example, in the media industry, ChatGPT can help news media workers intelligently generate reports, automate some labor-intensive editorial work, and generate content faster, more accurately, and more intelligently, thus improving the timeliness of news. This AI application has been around for some time. In March 2014, the *Los Angeles Times* website's robot reporter Quakebot wrote and published related information just three minutes after the earthquake in Los Angeles. The intelligent writing platform Wordsmith from the Associated Press, can write 2,000 articles per second. China Earthquake Network's writing robot compiled relevant information in just seven seconds after the Jiuzhaigou earthquake. The First Financial Network's "DT Draft King" can write 1,680 words in one minute. However, the emergence of ChatGPT has further promoted the integration of AI and media.

In the film and television industry, ChatGPT customizes film and television content based on the interests of the public, which is more likely to attract the public's attention and achieve better ratings, box office, and reputation. On the one hand, ChatGPT can provide new ideas for script creation, and creators then screen and rework the generated content from ChatGPT to stimulate their inspiration, expand their creative ideas, and shorten the creative cycle. On the other hand, ChatGPT has the advantages of cost reduction and efficiency improvement, which effectively helps the film and television production team reduce the cost of content creation, improve the efficiency of content creation, and produce higher-quality film and television content in a shorter time.

In 2016, New York University used AI to write the script for "Sunspring," which was then produced and ranked in the top ten in the London Sci-Fi Film Festival's 48-hour competition. Haima Qingfan Technology, a Chinese company, has launched an intelligent writing function for "novel to script" that has served more than 30,000 episodes of TV series scripts, more than 8,000 movie/network movie scripts, and more than 5 million online novels, including the popular works *Hi, Mom* and *The Wandering Earth*. In 2020, students from Chapman University in the US used the previous generation of ChatGPT, the GPT-3 model, to write a script and produce a short film called *The Attorney*.

Regarding marketing, ChatGPT creates virtual customer service to empower product sales. ChatGPT virtual customer service provides customers with 24/7 product recommendations and online services while reducing business marketing costs and promoting rapid growth in marketing performance. Additionally, ChatGPT virtual customer service quickly understands customer needs and pain points, bringing businesses closer to their consumer base and shaping a brand image that follows technology trends and is youthful. In situations where human customer service is limited and of varying quality, ChatGPT virtual customer service is more stable and reliable, with greater control over the brand image and service attitude compared to human customer service.

Today, substituting human intelligence for AI is still in continuous learning and development and presents a trend toward specialized research within different fields. Once AI takes over human professional abilities and achieves universal capabilities across different fields, it will undoubtedly become a "human-like" entity, completely opening up people's imagination about AIGC and ushering in the true era of AIGC.

5.3 ChatGPT's Impact on Medical Field

Currently, the widespread use of AI in the healthcare field is forming a global consensus. In fact, AI projects have been present in healthcare for quite some time, with AI-assisted diagnosis and AI-assisted decision-making in medical imaging gradually entering clinical practice. It can be said that AI is uniquely defending human health and well-being. The emergence of ChatGPT further accelerates the implementation of AI in the medical field and demonstrates exciting application prospects.

5.3.1 *The Reliability of ChatGPT and Doctors*

The US Medical Licensing Examination is known for its difficulty. Still, US researchers have found that the ChatGPT chatbot only passes or comes close to passing the exam with special training or reinforcement learning.

Researchers participating in the study are mainly from the US healthcare startup AnsibleHealth. They excluded image-based questions from the 376 exam questions released by the US Medical Licensing Examination website in June 2022 and asked ChatGPT to answer the remaining 350 questions. These questions are of various types, including open-ended questions that require candidates to diagnose patients based on existing information and multiple-choice questions that require candidates to determine the cause of the disease. Two evaluators were responsible for grading.

The results showed that after excluding ambiguous answers, ChatGPT scored between 52.4% and 75% in the three exam sections, and a score of around 60% is considered a passing score. Among them, 88.9% of ChatGPT's subjective answers included "at least one important insight," which means the insight is novel, clinically effective, and not obvious to everyone. Researchers believe that "achieving a passing score in this notoriously difficult professional exam and doing so without any human reinforcement (training) is a noteworthy event in the application of AI in clinical medicine," indicating that "large language models may have the potential to assist medical education and even clinical decision-making."

In addition to passing medical exams, ChatGPT's diagnostic level has also been recognized by the industry. The *Journal of the American Medical Association* (JAMA) published a research report discussing the rationality of using online dialogue AI models represented by ChatGPT in cardiovascular disease prevention advice, stating that ChatGPT has the potential to assist clinical work, help strengthen patient education, and reduce barriers and costs in communication between doctors and patients.

Based on current guidelines for CVD three-level prevention and healthcare recommendations and clinical doctor treatment experience, the researchers set up 25 specific questions involving disease prevention concepts, risk factor consultation, examination results, and medication consultation. Each question was asked to ChatGPT three times, and each reply was recorded. Each of the three replies for each question was evaluated by one reviewer, and the results were categorized as reasonable, unreasonable, or unreliable. If one of the three replies had an obvious medical error, it was directly judged as "unreasonable."

The results showed that ChatGPT's rational probability was 84% (21/25). The answers to these 25 questions show that the online dialogue AI model can answer CVD prevention questions, with the potential to assist clinical work, strengthen patient education, and reduce barriers and costs in communication between doctors and patients.

In fact, a large part of doctors' time globally is spent on various paperwork and administrative tasks, squeezing the time doctors spend on more important diagnoses and communication with patients. In a 2018 survey in the US, 70% of doctors said they spent more than 10 hours a week on paperwork and administrative tasks, and nearly one-third spent 20 hours or more.

An article published in the *Lancet* on February 6 by two doctors from the renowned St. Mary's Hospital in the UK pointed out that healthcare is an industry with a large, standardized space, especially in documents. We should respond to these technological advances. Among them, the "discharge summary" is considered a typical application of ChatGPT, as they are largely standardized in format. After doctors input brief explanations of specific information, concepts requiring detailed explanation, and instructions that need to be explained, ChatGPT outputs a formal discharge summary in a few seconds. Automating this process reduces junior doctors' workload and gives them more time to provide services to patients.

Of course, the current ChatGPT is not yet perfect for the medical industry, and there are also existing bugs—it provides inaccurate information, fabrication, and bias, which means that its application in this highly professional threshold industry should be more cautious. However, in any case, ChatGPT has opened up a new era of AI medical applications. On the one hand, it allows us to see that the era of Internet medicine will be accelerated by using ChatGPT for online consultations. Based on a powerful diagnosis and treatment database, as well as extensive training in the latest medical knowledge, ChatGPT can provide more professional and objective diagnostic advice than an average doctor. It can also achieve a real-time multi-user synchronous diagnosis.

For example, at the 17th European Crohn's and Colitis Organization (ECCO 2022) held in 2022, the discussion topic on endoscopy and histopathology was explored and clarified by medical experts worldwide. The more important impact of this meeting was that it proposed that under the trend of medicine + AI, AI reading of endoscopy and histopathology has become an important direction for development. At this meeting,

French medical expert Laurent Peyrin-Biroulet introduced a study using AI to read the histological disease activity of UC.

This study used histological images of 200 UC patients from the database of Vandoeuvre-lès-Nancy Hospital in France and entered them into an AI system that can self-judge histological progression and calculate the NANCY index (a validated histological index composed of "ulceration, acute inflammatory cell infiltration, and chronic inflammatory cell infiltration" with five disease activity levels defined as grades 0–4). In simple terms, the hospital used histological images of 200 UC patients, combined them with their developed AI reading system for diagnosis, and then compared the system's reading results with the reading results of three human pathologists, using the intra-group correlation coefficient (ICC) to understand and validate the feasibility of AI reading for UC diagnosis and treatment.

The results showed that the average ICC among the three pathologists was 89.33, while the average ICC between manual and AI readings was 87.20. From the comparison results, the AI readings were very close to the manual readings. And this is based on an AI system trained on a small sample size. With more training data, the accuracy of AI readings will far exceed the level of human experts, and the efficiency of the readings will also surpass human experts.

For example, Professor Zhang Lanjun, a renowned thoracic surgeon and chief expert on lung cancer at Sun Yat-sen University Cancer Center in China, partnered with Tencent in 2018 to develop an advanced image recognition system using neural convolutional algorithms to accurately identify pulmonary nodules based on diagnostic experience and features of benign and malignant nodules. With the continuous addition of data and training, the AI system could accurately identify pulmonary nodules. Zhang competed with the AI system and the hospital's senior experts to compare their diagnostic abilities. The results showed that the AI system was not inferior to the senior doctors in terms of accuracy, demonstrating that machines can diagnose based on rules or diagnostic standards and are not affected by human factors, resulting in lower error rates than human doctors.

On the other hand, ChatGPT also disrupts the medical industry and will be very effective in solving the current differences in medical standards and the problem of difficult access to medical care. According to data from the World Health Organization, it is estimated that by 2030, there will be a shortage of 10 million medical personnel globally, mainly in low-income countries. *Forbes* pointed out in an article on February

6 that AI can expand people's access to high-quality healthcare in areas where medical services are scarce. In the future, AI doctors can replace most routine diagnoses of illnesses. The disruption brought by AI to the medical industry has already begun. In the future, we may be more willing to accept the diagnosis of AI doctors than actual doctors. Time may tell us whether we prefer machines over humans in the face of rigorous and rule-based technologies, and perhaps AI will be more reliable than humans.

5.3.2 Digital Therapy

Today, if we need medical treatment, the traditional way is to use medication and medical equipment as the main treatment method. Imagine one day we go to the hospital for treatment, and the doctor prescribes software instead of medication, telling us to "remember to play for 15 minutes every day." This seemingly incomprehensible scene may soon become a reality in the clinic thanks to a new treatment method based on digital technology—digital therapy, with AI as a key driver for its clinical application and popularization.

First, let's take a look at what digital therapy is. The concept of digital therapy became popular in the US as early as 2012. According to the official definition of the US Digital Therapy Alliance, digital therapy is a software-based, evidence-based intervention that is used to treat, manage, or prevent diseases. With digital therapy, patients receive evidence-based treatment and prevention and manage their physical, psychological, and disease conditions. Digital therapy can be used independently or with drugs, equipment, or other therapies.

Simply put, in traditional treatment, patients often go to the pharmacy to obtain medication based on the prescription given by the doctor. Digital therapy replaces medication with a specific mobile app or a combination of software and hardware. Digital therapy could be a game or a behavioral guidance program that uses behavior intervention to induce cellular or even molecular changes, thus affecting disease conditions.

For example, if we go to the doctor for chronic insomnia, there are two traditional treatment methods: one is for the doctor to prescribe drugs like sedatives; the other is cognitive behavioral therapy for insomnia (CBT-I), which requires the doctor to provide face-to-face treatment. However, this clinical non-drug intervention is limited by the limited number of doctors and time and space constraints, which reduces its efficacy.

At this point, if the doctor prescribes a digital therapy like Somryst ® certified by the US Food and Drug Administration (FDA), it is equivalent to moving the off-line cognitive behavioral therapy online, freeing it from the constraints of doctors and time and space limitations. Somryst ® includes a sleep log and six guidance modules. Patients complete the six modules in order, recording their sleep situation each day and completing about 40 minutes of the course. Different stages have different courses; eventually, patients develop good sleep habits over a nine-week treatment period.

The greatest significance of digital therapy lies not in technological breakthroughs but in its innovation in the form of medication. This form has also updated people's understanding of disease treatment and improved disease treatment methods. Digital therapy is currently the most widely used in mental illness. It effectively treats depression, attention deficit hyperactivity disorder (ADHD), cognitive impairment in the elderly, and schizophrenia. In the process of application, AI plays a key role.

Looking specifically at the medical field, no reliable biomarkers are available for diagnosing mental illness. Psychiatrists have struggled to find shortcuts to identifying negative symptoms, which are symptoms that healthy individuals do not experience. These symptoms are not as obvious as the so-called "positive" symptoms, which indicate the presence of additional symptoms, such as hallucinations. One of the most common negative symptoms is a speech or language disorder. Patients will often speak as little as possible and frequently use vague, repetitive, and stereotyped phrases, which psychiatrists call "low semantic density."

Low semantic density is a warning sign that an individual may be at risk of developing a mental illness. Some research projects have shown that people at high-risk of developing a mental illness tend to use fewer possessive pronouns, such as "my," "his," or "our." Based on this, researchers have turned to ML to identify patterns in semantic usage as a breakthrough in diagnosing mental illness.

Today, the Internet has deeply integrated into society and people's lives. Ubiquitous smartphones and social media have made people's language more easily recordable, digitizable, and analyzable than ever before. Suppose ChatGPT could analyze data such as language choices, sleep patterns, and frequency of calls to friends. In that case, it could more closely and continuously measure various biological features of a patient's daily life, such as emotions, activities, heart rate, and sleep, and correlate this information with clinical symptoms to improve clinical practice.

Another concrete example comes from researchers using ChatGPT to diagnose Alzheimer's disease (AD). In a paper published on *PLOS Digital Health* on December 22, 2022, two scholars from Drexel University in the US used ChatGPT for diagnosing AD. As the most common form of dementia, Alzheimer's disease is a degenerative central nervous system disease, and scientists have been researching for years to develop effective drugs for AD, with limited progress so far.

The diagnosis of AD usually involves reviewing medical history and lengthy physical and neurological assessments and tests. As 60%–80% of dementia patients have language impairments, researchers have been focusing on applications that can capture subtle language clues, including identifying hesitation, grammatical and pronunciation errors, and word forgetfulness, as a fast and low-cost means of screening early-stage AD. The Drexel University study found that OpenAI's GPT-3 program can recognize clues from spontaneous speech and predict early-stage dementia with 80% accuracy. AI can be used as an effective decision support system, providing valuable data for doctors to use in diagnosis and treatment. The human eye may miss small anomalies in CT scans, but a trained AI can track the smallest details. After all, every doctor's memory has limitations that cannot compete with the powerful storage of a computer.

5.3.3 ChatGPT and Medicine Creating

In addition to its role in medical consultation and digital therapy, ChatGPT has the potential to revolutionize disease and drug research, which is where its strengths lie. The pharmaceutical industry is a dangerous and fascinating industry that is expensive and time-consuming. Typically, drug development can be divided into two stages: drug discovery and clinical research.

In the drug discovery stage, scientists must establish a disease hypothesis, identify targets, design compounds, and conduct preclinical studies. Traditional pharmaceutical companies must perform many simulated tests during the drug development process, which is lengthy, costly, and has a low success rate. According to data from *Nature*, the cost of developing a new drug is about $2.6 billion, takes about 10 years, and has a success rate of less than 10%. Finding a target, the drug's binding site in the body, often requires constant experimental screening to find a chemical molecule that has therapeutic effects from hundreds of molecules.

In addition, human thinking has a certain convergence. For new drugs targeting the same target, avoiding similar structures and even triggering patent litigation is sometimes difficult. Finally, a drug may require screening of thousands of compounds; even so, only a few can smoothly enter the final development stage. From 1980 to 2006, despite annual investments of more than $30 billion, on average, researchers could only find five new drugs each year. The key issue here is the complexity of discovering targets.

It is worth noting that most potential drug targets are proteins, and the structure of proteins, that is, the way 2D amino acid sequences fold into 3D proteins, determines their function. A protein with only 100 amino acids is already minimal, but even with such a small protein, the number of possible shapes it can produce is astronomical. This is also why protein folding has always been considered a problem that even large supercomputers cannot solve.

However, AI can determine the possible distances between amino acid pairs and their chemical bonds by mining large datasets—which is the basis for protein folding.

Flagship Pioneering, a very famous venture capital firm in the life sciences field, is acclaimed for incubating Moderna, and its founder, Dr. Noubar Afeyan, a PhD in bioengineering from MIT, wrote in his outlook for 2023 that AI will change biology in this century, just as bioinformatics changed biology in the last century.

Dr. Afeyan pointed out that advances in ML models, computing power, and data availability solve previously unresolved challenges and create opportunities for developing new proteins and other biomolecules. In 2023, his team at Generate Biomedicines demonstrated that these new tools could predict, design, and ultimately generate entirely new proteins whose structures and folding patterns are reverse-engineered to encode the desired therapeutic function.

When drug development enters the clinical research stage, which is the most time-consuming and expensive stage in the entire drug approval process, clinical trials are conducted in multiple stages, including phase I (safety), phase II (efficacy), and phase III (large-scale safety and efficacy) testing. In traditional clinical trials, the cost of recruiting patients is high, and information asymmetry is the primary problem that needs to be solved. According to a survey by CB Insights, the biggest reason for the delay in clinical trials is the recruitment of personnel, with about 80% of trials unable to find ideal drug volunteers on time. However, this problem is easily solved by AI. For example, AI uses technical means to extract useful information from patient medical records and match them with ongoing clinical studies, greatly simplifying recruitment.

For the experiment, there are often problems, such as the inability to monitor patient drug compliance. AI technology can achieve continuous monitoring of patients, such as tracking drug intake through sensors and patient medication compliance using images and facial recognition. Apple has launched the open-source frameworks ResearchKit and CareKit, which helps recruit patients for clinical trials and help researchers remotely monitor patients' health status and daily life using applications.

Although ChatGPT is imperfect and still has bugs, its disruptive power must be maintained. ChatGPT has a learning ability equal to humans based on learning from massive amounts of data. Given time, ChatGPT may assist doctors in their clinical work, enhance patient education, further promote the development of digital therapy, and assist in drug development. Especially for developing targeted drugs, the involvement of AI technology will greatly accelerate the process and reduce costs. In the future, although ChatGPT may not completely replace doctors, the future of medicine will certainly be collaborative between humans and machines.

5.4 ChatGPT's Impact on Legal Field

ChatGPT will be closely integrated with real-world applications and significantly impact various industries. Even the elite law profession has been affected by ChatGPT, and its impact on lawyers' practice and judges' judgments is gradually unfolding. It can be said that a technological revolution in the legal industry is coming.

5.4.1 AI and Legal Industry

Lawyers have always been regarded as an "elite" profession, with strong professionalism and dealing with complex cases and issues. Moreover, the litigation process in which lawyers participate directly affects the court's sentencing results, making their role particularly important. However, lawyers often face complex work and heavy pressure behind this "elite," professional, and important role. As the Internet says, "Being a lawyer is exchanging time for money"—the rhythm of 996 is not only the norm for programmers but also for lawyers.

Lawyers are usually divided into litigation lawyers and non-litigation lawyers. Simply put, litigation lawyers accept the client's commission to help them fight lawsuits. In addition to defending in court, the preliminary work of litigation lawyers also includes reading case files, drafting documents, collecting evidence, and researching legal materials. Some major cases may involve dozens to hundreds of case files. Non-litigation lawyers do not appear in court and are responsible for checking various materials, modifying various documents, and producing various legal opinions, agreements, etc. It can be said that whether it is a litigation lawyer or a non-litigation lawyer, a large part of their time is spent on desk work, dealing with a large number of documents, information, and contracts. However, the rigor of the law also requires lawyers to be meticulous. But it is precisely this kind of repetitive, mechanical work that is the corresponding advantage of AI.

The combination of AI and law can be traced back to the expert systems that started in the mid-1980s. The first practical application of expert systems in law was the Legal Decision Support System (LDS) developed by D. Wortman and M. Peterson in 1981. At that time, researchers regarded it as a practical tool for legal application. They tested a certain aspect of the US civil law system, using strict liability, comparative negligence, and damages models to calculate the compensation value of liability cases, successfully introducing the development of AI into the legal industry.

Since then, expert systems in law have begun to play an important role in the auxiliary retrieval of regulations and precedents, freeing lawyers from some mental labor. Obviously, lawyers would consume a lot of energy and time without computer compilation, classification, and retrieval of the vast amount of case files.

Moreover, due to the human brain's limited cognitive and memory abilities, there are also problems of incomplete retrieval and inaccurate memory. AI legal systems have powerful memory and retrieval functions, which can make up for certain limitations of human intelligence, help lawyers and judges engage in relatively simple legal retrieval work, and greatly free their mental labor to focus on more complex legal reasoning activities.

ChatGPT is significantly impacting law and is gradually changing the role of lawyers and judges. The legal profession has always been considered an elite profession with strong expertise, dealing with complex cases and issues that directly affect the outcome of court judgments. However, lawyers often face tedious work and heavy pressure. The use of AI can be of great advantage in this regard.

In terms of legal consultation, the first robot lawyer, Ross, provided immediate answers to clients' legal questions and personalized services in 2016. Ross solves problems like practicing lawyers typically do by first understanding the problem itself and breaking it down into legal issues, conducting legal research, and summarizing knowledge and experience to provide solutions. Unlike human lawyers, however, Ross completes the task relatively quickly.

In contract drafting and review services, AI can learn to generate highly refined and complex contracts suitable for specific situations by studying massive amounts of contracts. AI assembles contract clauses based on different scenarios to provide drafting services for basic contracts and legal documents. For example, in the case of a sales contract, answering a series of questions posed by the AI program, such as the subject matter, price, delivery location, method, and risk transfer, would result in a complete initial draft of the sales contract, which may even be superior to the results of many experienced legal advisers.

5.4.2 ChatGPT's Success in Judicial Exam

The integration of AI into the legal industry is already a reality, and the emergence of ChatGPT has once again highlighted the rapid development of AI technology. ChatGPT has passed the bar exam, and AI lawyers are almost within reach.

Specifically, the Uniform Bar Examination in most US states has three components: multiple-choice questions (MBE), essays, and performance tests. The multiple-choice questions consist of 200 questions from 8 categories and usually account for 50% of the total score. Based on this, researchers evaluated the performance of OpenAI's text-DaVinci-003 model (commonly known as GPT-3.5, which is the public version of ChatGPT) in the MBE.

Researchers purchased official standard exam preparation materials, including practice questions and simulated exams, to test the actual effect. The text of each question was automatically extracted, with four multiple-choice options and the answers stored separately, consisting only of the correct letter answer for each question, without explanations for the correct and incorrect answers. Subsequently, the researchers attempted to prompt engineering, hyperparameter optimization, and fine-tuning for GPT-3.5. The results showed that hyperparameter optimization and prompt engineering positively impacted GPT-3.5's performance, while fine-tuning had no effect.

Finally, in the complete MBE practice exam, an average accuracy rate of 50.3% was achieved, greatly exceeding the baseline guessing rate of 25% and reaching an average pass rate in evidence and torts. Especially in the evidence category, it is on par with human performance, maintaining a 63% accuracy rate. GPT lagged behind human test takers in all categories by about 17% on average. In the cases of evidence, torts, and civil litigation, this gap can be negligible or only in the single digits. Overall, this result greatly exceeded the researchers' expectations. This also confirms ChatGPT's general understanding of the legal field rather than random guessing.

Moreover, in the admission exam of the Florida A&M University College of Law, ChatGPT also scored 149 points, ranking in the top 40%. Reading comprehension questions performed the best.

Currently, although ChatGPT cannot completely replace human lawyers, AI represented by ChatGPT is rapidly entering the legal industry. The application of technological achievements to legal services has become an indisputable fact, and AI technology is deeply affecting the future direction of the legal service industry and market.

On the one hand, from a "beneficial" perspective, if ChatGPT is used well, lawyers can leave work early. In the foreseeable future, with ChatGPT continuously fed with large amounts of professional data from the legal industry, it will easily handle brief legal service work. If a lawyer needs to retrieve case law or legal provisions, they only need to input the keywords into ChatGPT. They immediately obtain the desired provisions and cases. ChatGPT provides initial opinions for basic contract review, and lawyers further refine and modify them. If a monetary calculation is needed, such as for compensation in a traffic accident or personal injury case, ChatGPT can also quickly provide data. In addition, for tasks such as proofreading and translating text, file classification, creating visual charts, and writing formatted briefs, ChatGPT can also easily handle them.

On the other hand, we need to face the fact that when ordinary legal services are replaced by AI, lawyers in corresponding positions will gradually exit the market, which will inevitably impact some lawyers' value and functional positioning. Obviously, compared to human lawyers, AI lawyers work more efficiently and effectively while requiring less labor cost. Therefore, their charging standards may be relatively lower.

In the future, with the involvement of ChatGPT, the supply and demand information of the legal services market will become more transparent, and the operation process and charging standards of online legal services products will become more open. In other words, AI's convenience, transparency, and controllability in providing legal services will become an advantage in attracting clients. Under such circumstances, the business expansion opportunities, personal growth rate, and professional moat construction for lawyers will be greatly affected.

It's important to note that the traditional legal service industry is a "people-oriented" industry, with both the service provider and recipient being human. When AI takes the lead in resolving simple cases in legal services, the legal service market will form a diversified phenomenon of service providers, and the work and functions of human lawyers will be redefined and evaluated. The business model of the legal services market will also change.

For the judiciary, a field with obvious rules and standards, the future AI-based judicial system will more effectively ensure the fairness and justice of the rule of law. AI judges will likely become a reality shortly.

5.5 ChatGPT's Impact on Education Field

In the many industries that ChatGPT is about to change and disrupt, education is one of the industries that has received a lot of attention. Humans always use tools to understand the world, and the invention and innovation of tools drive the progress of human history. Similarly, changes and innovations in educational methods and means also drive the progress and development of education.

5.5.1 ChatGPT for Homework

The fact that AI is changing education is an inevitable and happening reality. Before ChatGPT, many AI products had already been used in education, such as in various types of education, including early childhood education, higher education, vocational education, etc., AI had been applied in scenarios such as photo-based problem-solving, hierarchical scheduling, oral evaluation, paper setting and marking, composition correction, homework assignment, etc.

The explosion of ChatGPT further impacted education. One of the most direct manifestations is that students started using ChatGPT to complete their homework.

An anonymous survey by the Stanford University campus media, *Stanford Daily*, showed that about 17% of the surveyed Stanford students (4,497) said they used ChatGPT to assist them in completing autumn assignments and exams. Dee Mostofi, a spokesperson for Stanford University, said that the school's Judicial Affairs Committee has been monitoring emerging AI tools and will discuss how they relate to the school's honor code.

A survey by the online course provider, Study.com of 1,000 students aged 18 and above worldwide, showed that more than 9 out of every 10 students know about ChatGPT, more than 89% of students use ChatGPT to complete homework, 48% of students use ChatGPT to complete quizzes, 53% of students use ChatGPT to write papers, and 22% of students use ChatGPT to generate paper outlines.

The sudden arrival of ChatGPT has alerted the global education industry. As a result, schools in some regions of the US have had to ban ChatGPT altogether. Some have developed software specifically to check whether AI completes text assignments submitted by students. A spokesperson of New York City Department of Education believes the tool "will not foster critical thinking and problem-solving skills."

Philosopher and linguist Avram Noam Chomsky has even stated that ChatGPT is essentially "high-tech plagiarism" and "a way of avoiding learning." Chomsky believes students instinctively use high tech to avoid learning, which is "a sign of the failure of the education system."

Of course, while holding up the banner of opposition, there are also different voices and reflections on this. For example, Zhao Bin, a teacher at Fudan University, believes in the attitude of "if you can't beat them, join them." Zhao said that ChatGPT would become an essential tool in his teaching. In the first few classes of the new semester this year, he will tell his students to learn ChatGPT. According to Zhao's preliminary ideas, after the students finish the lesson, they will talk to ChatGPT, understand some new things, organize the content, and submit an assignment. As Zhao said, "Because what I am more concerned about now is the students' ability to ask questions, that is, what kind of questions they will ask the machine after class, what kind of knowledge they want to learn, this is my focus."

Any new technology, especially revolutionary technology, will inevitably spark controversy. For example, the advent of cars once sparked strong opposition from carriage drivers. Objectively speaking, the era of AI is an inevitable trend, and ChatGPT has brought the era of AI closer to us. It suddenly came, and it can really help us deal with work. Not only can it help us deal with work, but it can also do it better than humans. This will inevitably cause some people to oppose it. However, whether we are against it or choose to embrace it, it will not change the arrival of the era of AI.

In education, we do not need to worry about whether ChatGPT can help students complete homework or help students write papers. Especially for exam-oriented education, if we only cultivate children into a knowledge base and problem-solving machine, we completely compete with AI based on big data databases, which is a mistake.

Embracing ChatGPT and making it an auxiliary tool for student knowledge acquisition in teaching can liberate teachers from the burden of rote teaching and give them more time to think about cultivating innovative and creative thinking. In the face of the era of AI, if we continue to educate and train using standardized test questions and standardized answers, we will become the carriage drivers of the first industrial revolution.

The AI revolution triggered by ChatGPT poses new challenges for education in both China and the US, that is, how do we change our education with this powerful AI assistant to maintain human innovation and lead AI to help our society develop.

5.5.2 ChatGPT's Challenges to Teachers

The impact of ChatGPT on the education sector has sparked a debate on whether "ChatGPT will replace teachers." This question is two-sided and depends on how we define a teacher's role. In the age of AI, where knowledge acquisition is no longer difficult or scarce, traditional knowledge-based teaching that focuses on rote learning and delivering content will eventually be replaced by AI. In terms of knowledge acquisition, AI performs better than most teachers due to its ability to access and analyze vast amounts of data.

Moreover, AI can not only teach better but also learn better. Chinese students are widely regarded as the best test-takers globally. They have a vast knowledge base, solid foundational knowledge, and broad coverage. While this was considered an advantage in the past, it has become increasingly embarrassing in the face of ChatGPT.

In 2017, Qian Yingyi, a counselor of the State Council of China and dean of the School of Economics and Management at Tsinghua University, pointed out that China's education system has a systemic bias toward equating education with knowledge. He expressed concerns that "future AI will render the advantages of our education system irrelevant." This awkward point in time shows that today's so-called advantage of comprehensive knowledge can easily be replaced by ChatGPT, which is evolving as a digital brain.

However, suppose we redefine the role of a teacher to focus on fostering and exploring human-specific imagination, creativity, and inspiration. In that case, AI will need more work to replace them. In the age of AI, our competition with machines will not be at the knowledge level but at the level of our uniquely human qualities, such as imagination, creativity, and innovation.

In other words, what we need to do in education is to focus on cultivating and unleashing our unique human characteristics, such as creativity, imagination, and inspiration. Only by tapping into these unique qualities can we make AI a tool for achieving our dreams and not become tools for AI training.

Therefore, discussing whether ChatGPT will replace teachers is meaningless, and the key lies in our choices as humans. The arrival of the age of AI has shown us the urgency of education reform in China.

5.5.3 The Impact on Academic Field

In addition to its impact on traditional education, the ChatGPT wave has also affected the research and academic fields.

Within a week, the top international journal *Nature* published two articles discussing the impact of ChatGPT and AIGC on the academic field. *Nature* stated that since every author bears responsibility for the published work, and AI tools cannot do this, no AI tool will be accepted as a coauthor of research papers. The article also pointed out that if researchers use relevant programs, they should be explained in the methods or acknowledgments section.

Science directly banned submissions using ChatGPT-generated text. On January 26, *Science* announced through an editorial that it was updating its editorial rules, emphasizing that text, numbers, images, or graphics generated by ChatGPT (or any other AI tool) could not be used in works. The editorial emphasized that AI programs cannot be authors. Any violation would constitute scientific misconduct.

However, the trend is already apparent, and an undeniable fact is that AI can indeed improve efficiency in the academic circle.

On the one hand, ChatGPT can improve the efficiency of retrieving and integrating basic research materials, such as some review work, which AI handles quickly, and researchers can focus more on the experiments themselves. In fact, ChatGPT has become the digital assistant of many scholars. Casey Greene, a computational biologist, uses ChatGPT to revise papers. ChatGPT can review a manuscript in just five minutes, and even problems in the reference section can be identified. Neurobiologist Almira Osmanovic Thunström believes that language models help scholars write funding applications, saving scientists more time.

On the other hand, ChatGPT only does limited information integration and writing at the current stage and cannot replace in-depth and original research. Therefore, ChatGPT can motivate academic researchers to conduct more in-depth research.

Faced with the impact of ChatGPT in the academic field, we must admit that many jobs in the human world could be more effective. For example, when we cannot distinguish whether a machine or a human writes an article, these articles have no value. Now, ChatGPT is driving change and innovation in the academic world. ChatGPT breaks down formalistic texts, including various reports and most papers, and humans can use ChatGPT to create precious and contributory research.

ChatGPT may revolutionize the academic world, urging researchers to devote more time to truly thoughtful and constructive academic research rather than copying and pasting format papers.

5.6 ChatGPT's Impact on New Retail

Since the concept of New Retail emerged in 2016, various projects have emerged like mushrooms after rain. The emergence of New Retail is driven by technology. Now, as ChatGPT represents a huge breakthrough in AI technology, it also shows remarkable imagination space for the New Retail industry.

5.6.1 *Technological Support behind New Retail*

Looking back, China's retail industry has undergone a long development process, from traditional retail to Internet e-commerce, and has experienced ups and downs. Before the 1990s, the retail form was still physical and basically specialty stores. After that, in response to the situation, specialty stores were reorganized to form department stores. After the 1990s, chain supermarkets became mainstream in the retail market, while there were also modern specialty stores, specialty supermarkets, and convenience stores. At the same time, the competition among chain supermarkets became increasingly fierce, forcing the market to enter a period of integration.

Around 2000, large-scale comprehensive supermarkets and discount stores appeared, and foreign retail companies such as Carrefour entered the Chinese market, opening up a new battleground for the Chinese retail industry. After 2000, the number of supermarkets in the Chinese market skyrocketed, and shopping centers that combined retail and services also began to emerge and develop, moving toward comprehensive shopping centers that integrate entertainment, catering, services, shopping, and leisure, making the Chinese retail industry a colorful landscape.

But behind this colorful landscape, a huge threat is gradually approaching. The development of the Internet and e-commerce has dealt a serious blow to China's traditional retail industry. Many physical stores have closed, and some department stores have gone bankrupt. Around 2013, influenced by the mobile Internet, the retail industry and consumers' consumption habits and concepts were affected. The online retail industry was exceptionally hot during this period, while off-line stores were exceptionally bleak. Moreover, the focus of e-commerce shifted from the PC to the mobile end.

In 2015, e-commerce entered a stable development stage. Influenced by the "Internet+" and "O2O model," many off-line retail enterprises began to explore the path of integration and development with e-commerce. Since 2016, there has been a great fluctuation in China's retail industry, with large-scale off-line supermarkets closing down one after another, especially since the closure of Darunfa was shocking; the traffic dividend of pure e-commerce is gradually disappearing.

On October 13, 2016, Jack Ma said at the Alibaba Cloud Computing Conference, "The pure e-commerce era will soon come to an end. In the next ten to twenty years, there will be no such thing as e-commerce, only New Retail." What is New Retail? Jack Ma explains that true New Retail can be generated only by combining online, off-line, and logistics. Essentially, it uses digital and technological means to improve the efficiency of traditional retail.

2017 became the first year of China's New Retail development, with Internet giants led by Alibaba and Tencent investing heavily in off-line physical commercial areas, creating many new species such as Alibaba's Hema Fresh, JD's 7Fresh, Meituan's Palm Fish Fresh, and Yonghui's Super Species.

More and more industry giants have adopted the methodology of upgrading and transforming New Retail, which has become a major trend in the industry. As a New Retail representative, Hema Xiansheng is Alibaba's completely restructured off-line supermarket business model. Hema Xiansheng embodies all the characteristics of Alibaba's New Retail and has become the benchmark for Alibaba's New Retail. Consumers can purchase in-store or order on the Hema app. One of Hema's biggest features is fast delivery: within a three-kilometer radius of the store, goods can be delivered to the door within 30 minutes.

The emergence and development of New Retail must be connected to the promotion of technology. In the past decade, the wave of information technology has overturned the industrial ecological chain. New-generation information technologies such as cloud computing, big data, and AI have become important driving forces leading innovation in various fields. Technological progress has promoted the comprehensive transformation of the retail field infrastructure in the retail industry, developing the retail industry toward intelligence and coordination and ultimately achieving cost reduction and efficiency improvement.

In the process of retail transformation toward New Retail, AI technology is the main force. For example, by applying AI technology, businesses better understand consumer needs, improve service quality, and improve customer stickiness. In addition, through AI technology, products, and services can be made more accessible, competitive pricing strategies can be promoted, and the end-user experience can be improved. And the interaction with consumers and feedback information related to consumers can truly land with the help of ChatGPT. The advent of ChatGPT will further deepen the application of AI in New Retail, and a new business model centered on customers, consumer needs, and customized, personalized requirements will be opened with the help of ChatGPT technology, which will usher in a new change. It can be foreseen that the retail industry will still be a key application area for AI in the next few years.

5.6.2 ChatGPT's Benefits to New Retail

AI has penetrated various value chain links in the retail industry. The outbreak of ChatGPT will also promote the aggregation of AI applications in the retail industry from individual aspects.

ChatGPT can achieve personalized recommendations on the customer side, making it possible for merchants to adjust their products and promotion strategies quickly. Suppose a large amount of product knowledge is input and trained through algorithms for some time. In that case, ChatGPT's understanding of the product may be more professional than a ten-year salesperson because ChatGPT has a stronger memory and is better at choosing the best answer. With consumer data accumulation, merchants can adjust their product development and promotion strategies through ChatGPT. The more one understands customer behavior and trends; the more accurately one can meet consumer needs. In short, ChatGPT helps retailers improve demand prediction, makes pricing decisions, optimizes product placement, and ultimately connects customers with the right product at the right time and place.

In addition, ChatGPT helps improve supply chain management efficiency in the retail industry. One of the biggest challenges facing traditional retailers is maintaining accurate inventory. However, ChatGPT can connect the entire supply chain and consumer side, providing retailers with comprehensive and detailed data on stores, shoppers, and products, which helps retailers make more appropriate decisions on inventory management. In addition, ChatGPT can quickly identify out-of-stock items and pricing errors, reminding staff of insufficient or misplaced items to achieve timely inventory acquisition.

Furthermore, suppose ChatGPT is used for an online sales consultation. In that case, it can provide one-to-many service that is more professional than human customer service, which means that ChatGPT changes the problem of high human after-sales costs and low efficiency. Robot assistants will greatly improve the efficiency of the after-sales process. It can be foreseen that the future New Retail scene will be a highly contextualized and personalized shopping scene.

5.7 ChatGPT's Impact on Financial Industry

The ChatGPT craze is sweeping across various industries, including the financial sector. First, East Money Research released a more than 6,000-word medical aesthetics report written by ChatGPT, which went viral in the industry. Later, China Merchants Bank published a WeChat article titled "Family Credit Card Warmly Launched, ChatGPT's First Interpretation of 'Life's Reverse Journey, Family Affection is Priceless,'" intending to partner with ChatGPT in producing promotional materials. Of course, this is also the first attempt in the financial industry to collaborate with ChatGPT in creating promotional content. So, what kind of impact and influence will the rise of ChatGPT have on the financial industry?

5.7.1 AI and the Financial Sector

With the rise of ChatGPT, the finance industry has also begun to take notice of its potential impact and influence. Before ChatGPT, AI technology had already entered the finance industry.

According to a report by the China Academy of Information and Communications Technology on AI in the finance industry (2022), AI technology has penetrated all five major business chain links in the finance industry, including product design, marketing, risk control, customer service, and other supporting activities, and has covered mainstream business scenarios. Typical applications include intelligent marketing, identity recognition, and customer service. AI technology has entered the value creation stage, solved industry pain points, obtained incremental business, reduced risk costs, improved operating costs, and enhanced customer satisfaction.

In the front-end application scenarios, AI is changing how financial service enterprises obtain and maintain customers through intelligent marketing, intelligent customer service, and intelligent investment advisory services. Intelligent investment advisory services use AI algorithms to generate personalized asset allocation recommendations for users based on their risk preferences, financial situation, return objectives, and modern financial models such as investment portfolio theory. The portfolio is continually tracked and dynamically rebalanced.

Compared to traditional human investment advisers, intelligent investment advisory services have unique advantages: first, they provide efficient and convenient investment consulting services; second, they have low investment thresholds, low fees, and high transparency; third, they can overcome subjective emotional investment and achieve high investment objectivity and diversification; and fourth, they provide personalized wealth management services and rich, customized scenarios.

In investment, accurate, fast, and real-time data is the greatest value, precisely where ChatGPT excels. For example, in stock investment, ChatGPT captures various news and real-time monitoring of capital flows, combined with various technical analyses in the financial investment field, to provide a relatively objective analysis and recommendation. These investor recommendations are more objective, real-time, and comprehensive than human investment advisers.

AI applies to front-end work in financial investment and offers exciting changes for the middle and back offices. Intelligent investment has begun to be profitable and has enormous development potential. Some companies use AI technology to optimize algorithms, enhance computing power, and achieve more accurate investment predictions, increasing revenue and reducing tail risks. Through portfolio optimization, significant excess returns have been achieved in real-time trading. The development potential of intelligent investment in the future is enormous.

Intelligent credit assessment has advantages such as online real-time operation, automatic system judgment, and short audit cycles, providing a more efficient service model for small and micro-loans, and has already been widely used in some Internet banks.

Intelligent risk control is applied to credit evaluation, risk pricing, and collection in banking enterprise credit, Internet finance assistance, and consumer finance scenarios, providing a new online-based risk control model for the finance industry.

Although AI applications in the finance industry are still in the "shallow application" stage and mainly focus on the intelligent transformation of processes and repetitive tasks, the development potential of AI technology in the finance industry has been demonstrated as it penetrates the periphery and reaches the core of the financial business. The progress of AI technology will inevitably bring about complete automation of customer financial life in the future.

5.7.2 *Igniting Intelligent Finance with ChatGPT*

If current AI applications in finance are still in the early stages of "shallow application," the emergence of ChatGPT is like adding fuel to the fire for AI in the finance industry.

First, finance investment decisions depend highly on quantifiable and reliable information, such as data, history, industry policies, and development trends. AI's ability to acquire, integrate, and analyze such information is far superior to humans.

Second, ChatGPT simulates human chat behavior very well and performs better in understanding and interaction, pushing financial institutions toward more humanized services. On the one hand, ChatGPT generates natural language responses to meet customers' personalized consultation needs. Semantic analysis can identify customer emotions to understand their needs better and provide better service, significantly improving the accuracy and satisfaction of intelligent customer service and enhancing brand image. On the other hand, ChatGPT can assist financial institutions in forming enterprise-level intelligent customer service capabilities. Usually, B-end users have high professional thresholds and complex business scenarios. ChatGPT is expected to use deep learning technology to improve B-end user service efficiency and professionalism in such cases.

Moreover, ChatGPT's application in the finance industry can greatly improve work efficiency and bring business transformation. ChatGPT can help financial institutions quickly extract valuable key information from massive data, such as industry trends, financial data, and public opinion trends, and convert them into readable natural language texts, such as industry research reports, risk analysis reports, etc., greatly saving manpower costs.

For example, East Money has already used ChatGPT to write a securities research report called "Improving External Beauty and Enhancing Internal Confidence-The Revolution of Medical Beauty." The report is more than 6,000 words long. It includes sections such as an introduction to the medical beauty industry, an overview of the global medical beauty market, the rise of light medical beauty, the rise of medical beauty in China, major participants in the global medical beauty industry, and ChatGPT's views on the Chinese and global medical beauty market after the epidemic. For ChatGPT, these labor-intensive research report writing tasks are almost effortless.

Overall, ChatGPT's prospects in the finance industry are promising, and perhaps soon, ChatGPT will enable us to see the transformation of the finance industry.

The changes brought about by ChatGPT are far more than those discussed above, and it can be said that all jobs with rules and regulations in human society will be replaced. From education, healthcare, finance, law, manufacturing, management, media, publishing, and research to design industries, all existing divisions of labor in human society will undergo enormous changes due to AI intervention. In the future, there may be only two types of jobs left for humans: one is to lead AI, and AI may lead the other.

In this era of change, in an upcoming era of human-machine collaboration, just as humans have constantly adjusted their roles in the billions of years of natural evolution, the advancement of AI technology will drive us, humans, to adjust our roles once again and will inevitably trigger a new round of industrial restructuring and division of labor worldwide.

5.7.3 Challenge from Bloomberg

On March 30, 2023, according to the latest report released by Bloomberg, they constructed the largest specialized data set. They trained a language model specifically for the finance industry—BloombergGPT—with 500 billion parameters. The model is built on Bloomberg's vast financial data sources, creating a data set with 363 billion tags that support various tasks within the finance industry. The model outperforms existing models in financial tasks and performs comparably well in general scenarios.

The report indicates that researchers used Bloomberg's existing data to create, collect and organize resources and constructed the largest specialized data set to complete BloombergGPT. The model was trained using a combination of general and financial scenarios. Bloomberg is primarily a financial data company, and data analysts have collected many financial documents over the forty years since its establishment, covering a range of topics.

The developers added this data to a public data set to create a large training corpus with over 700 billion tags. Using part of this training corpus, they trained a model with Bloomberg style, 500 billion parameters, designed based on the guidance of Hoffman and Le Scao, and trained with a combination of general and financial scenarios.

The results show that our mixed training method significantly outperforms existing models in financial tasks and performs comparably or even better in general scenarios.

This means that with the technology of ChatGPT and Bloomberg's powerful financial data and information, Bloomberg will challenge the current investment banking and wealth management industries. Perhaps shortly, Bloomberg will become one of the largest investment banks in the age of AI.

5.8 ChatGPT's Impact on Humanoid Robots

As a mechanical device that automatically performs work, humanoid robots have significantly improved their level of intelligence in recent years with the application of AI interactive technology and have gradually entered the application stage. Now, the globally acclaimed ChatGPT has added fuel to the fire for humanoid robots, and perhaps with the arrival of ChatGPT, humanoid robots will usher in a new turning point in their development.

5.8.1 *The Human Dream of Robots*

Creating a humanoid robot that resembles humans has always been a human dream. So why create a robot that looks like a human?

There are mainly two reasons. First, it better serves as a human workforce. Elon Musk has repeatedly emphasized that one of human civilization's greatest risks is the labor shortage and that humans should focus their efforts on intellectual rather than physical labor. However, to enable robots to serve as a more efficient human workforce, they must be adaptable to human life. Because our society is designed based on humans themselves, a humanoid robot can better meet this condition. Humanoid robots can only be born in response to our society to achieve the highest efficiency as a workforce.

Second, it is demand-driven. In many fields, robots serve as attendants, and only a human appearance is more readily accepted. For example, in postpartum care, child companionship, and elderly care—human-like robots are easier to emotionally connect with humans, which is the first part of the "uncanny valley" effect. The "uncanny valley" effect was proposed by the father of modern Japanese simulation robots, Masahiro Mori, in 1970. He observed that when the appearance and movement of a simulation robot are similar but not perfectly matched, human observers will produce a disgusted reaction.

For example, our liking for humanoid robots or dolls increases as their level of simulation increases. When the level of simulation reaches a certain proportion, when we see a simulation robot that is neither human-like nor typical robot-like, our emotional response suddenly reverses, instinctively feeling abnormal and producing avoidance reactions such as disgust and fear. Only when the level of simulation continues to increase our emotional response will reverse again.

According to different application scenarios, humanoid robots can be roughly divided into industrial and service robots. The difference between industrial robots and service robots lies mainly in their different application areas—industrial robots are mainly used in the industrial production field. In contrast, service robots have a wider range of applications, including various aspects of social life.

In the industrial age, the demand for automation in manufacturing industries such as automobiles, electronics, and home appliances has driven the vigorous development of industrial robots. With the rise of the tertiary industry, the demand for automation in service industries such as medical, logistics, and catering is expected to drive the demand for corresponding service robot categories. Especially in high-risk service industries, such as medical care, rescue, and firefighting, the demand for replacing humans with machines is even stronger.

From the perspective of assisting humans, service robots can effectively improve the existing work efficiency of humans through motion control, human-machine interaction, and other technologies. These robots do not replace humans but coexist in a collaborative form.

For example, as the pace of life accelerates, people hope to be freed from tedious housework. The emergence of domestic robots makes people's lives more convenient and satisfies their pursuit of high-quality life.

From a simple tool-based application to emotional communication and daily companionship, service robots are gradually becoming a part of people's daily lives. It is worth mentioning the emergence of sex robots—in 2018, the world's first AI robot, Harmony, was officially launched, with a price of 7,775 pounds, equivalent to about ¥68,000.

It can be said that Harmony is a high-level expression of sex robots. Just from its appearance, you can see the designer's painstaking efforts. Harmony has over 30 different faces, from black to Asian faces, and to pursue perfection, designers will personally polish these faces, even spraying on a few freckles. In addition to highly simulating the softness and appearance of the human body, Harmony can also learn through AI, generate emotions with humans, and has 12 different personalities, such as kindness, sexiness, and innocence. Any Harmony model has its own exclusive app, which connects to the Internet and communicate with people using a corpus.

Of course, AI technology at that time needed to be more mature, and the high price also made consumers hesitate. But soon, with the advancement of technology, Harmony 2.0, an upgraded version of the first-generation robot, was born. Compared to the first-generation, Harmony 2.0 is more like a real partner. The facial expressions of Harmony 2.0 are more colorful, the limbs are more flexible, and the body skin is more realistic. Due to the internal heater, Harmony 2.0 can also simulate real body temperature. In addition, Harmony 2.0 integrates Amazon's Alexa voice system, which analyzes voice information and quickly provides accurate feedback on various topics. At the same time, Harmony 2.0 also has an intelligent software system that stores and remembers past chat content to determine the partner's habits and preferences. That is to say, after long-term companionship, Harmony 2.0 will better understand its partner.

For creating new fields, as the industry develops, service robots are also starting to explore "what humans cannot do" and "what humans do not want to do" to create new needs. Some specialized robots are used in special fields, such as extreme environments and precision operations, such as the Da Vinci surgical robot, anti-terrorism, and riot control robots, military drones, etc.

Among them, the Da Vinci surgical robot assists doctors in surgery and completes extremely delicate actions that human hands cannot do. The incision can also be tiny, speeding up the patient's postoperative recovery. Anti-terrorism and riot control robots replace people in exploring, eliminating, or destroying explosives in dangerous, harsh, and harmful environments. In addition, they can also be used in firefighting, rescuing hostages, and confronting terrorists. Military drones are used in reconnaissance and warning, tracking and positioning, special operations, precision guidance, information warfare, battlefield search and rescue, and other strategic and tactical tasks. They have been widely used in modern military fields.

5.8.2 ChatGPT + Humanoid Robots

Whether we accept it, living and collaborating with robots will be a routine mode of future society. This is also why tech giants are entering this industry. For example,

Dyson, which has "gone beyond" household appliances, has entered the field of humanoid robots. Currently, Dyson has released a mechanical arm that can pick up bleach and pick up dishes. Dyson's vision is to launch a robot that can do housework in the next 10 years. With experience and technology accumulated in household service fields such as vacuum cleaners, hair dryers, and vacuum cleaners, Dyson plans to use its advantage technology to create a home-made nanny humanoid robot.

Another example is carmaker Tesla, which unveiled its humanoid robot in 2022. Musk stated that the initial positioning of the Tesla robot was to replace people to do repetitive, boring, and dangerous work. Still, the long-term goal is to make it serve daily work for thousands of households. In addition, there are pure robotics companies represented by UBTECH and Boston Dynamics.

Although, in recent years, the production and research community's actions regarding humanoid robots have increased significantly, there has always been a lack of a breakthrough to promote the development of humanoid robots into the next stage. Humanoid robots are expensive, and the product experience is often unsatisfactory.

On the one hand, one of the major challenges for current humanoid robots is flexibility at the hardware level. Because robots are assembled from mechanical components that differ greatly from human skeletal and neural control systems, so much work still needs to be done at the hardware level to achieve human-like flexibility or at least to make humanoid robots appear more human-like.

On the other hand, current humanoid robots can only respond to standardized programming, with little to no intelligence beyond that. When faced with non-standardized problems, AI quickly becomes unintelligent, effectively becoming an "idiot." AI is mainly used for statistical and analytical work, including rule-based reading, listening, and writing tasks. It lacks logic and the ability to think critically and is still in the early stages of controlling an entirely robotic body. Human neural control systems have evolved over tens of thousands of years of training, so current humanoid robots are still in their infancy in terms of both AI and hardware control.

However, the emergence of ChatGPT could be a game-changer for developing humanoid robots. AI is seen as having achieved a qualitative breakthrough since AlphaGo, and with the growing strength of AI, humanoid robots may soon see an acceleration in their development and application. ChatGPT's large model technology can unlock new scenarios for humanoid robots, and combining the two could further enhance the robots' intelligence. Researchers have already found that ChatGPT has the mental age of a nine-year-old child based on a mental test. From a theoretical perspective, human and AI are just two sets of intelligence in this world. Both complex "algorithms" induce, learn, and reconstruct external world features from limited input signals. Therefore, in theory, as long as we continue to educate and train AI with massive amounts of data, AI will eventually be able to operate algorithms called "self-awareness." It is not surprising that AI can pass a mental test; However, ChatGPT currently only has the mental age of a nine-year-old; with more data training in the future, AI will have the ability to think and reason similar to humans.

The application of humanoid robots will expand from education and entertainment to fields such as healthcare for the elderly, disinfection, and logistics. The shift from automation to autonomous intelligence will bring significant development opportunities.

6

Are Humans Ready?

6.1 The Imperfect ChatGPT

Despite ChatGPT's unprecedented intelligence and charm, the objective fact is that ChatGPT's human-like output and remarkable generality are only the results of outstanding technology, not true intelligence. ChatGPT also has bugs and is not perfect.

6.1.1 ChatGPT's Disadvantages

One of the biggest criticisms of ChatGPT is its accuracy issues. Whether it's the previous generation GPT-3 or the current ChatGPT, both make some ridiculous errors, an inherent drawback of this method.

Because ChatGPT essentially generates data by maximizing probabilities rather than generating responses through logical reasoning, ChatGPT's training uses unprecedentedly large data. It is trained through AI models such as deep neural networks, self-supervised learning, reinforcement learning, and prompt learning. The number of model parameters for the previous generation GPT-3 disclosed is as high as 175 billion. ChatGPT can only demonstrate statistical association capabilities, can discern the correlations between words-words, sentence-sentences, and other relationships in massive data, and embodies the ability to engage in language dialogue, thanks to the combination of big data, big models, and big computing power. Because ChatGPT is trained on the standard of "co-occurrence means correlation," false associations and synthetic results of piecemeal information occur. Many ridiculous errors are due to mechanically matching data without common sense.

In other words, although ChatGPT synthesizes language answers through statistical relationships between the words it mines, it cannot judge the credibility of the contents in the answer, which leads to erroneous answers that may harm society, including biased answers, spreading toxic information that is inconsistent with facts, offensive or ethically risky content, etc. Moreover, if someone maliciously feeds ChatGPT with misleading or erroneous information, it will interfere with the knowledge generation results of ChatGPT, increasing the probability of misguidance.

We can imagine an intelligent machine that costs almost nothing to create, has an accuracy rate of about 80%, and can deceive non-professionals almost 100% of the time. With output speeds millions of times faster than human authors, this machine takes over the compilation of all encyclopedias and answers all knowledge questions. This poses a huge challenge to human knowledge and memory that relies on the brain.

For example, ChatGPT needs to be fed with more relevant data in life sciences to generate appropriate answers and may even produce nonsensical responses. In life sciences, there are higher requirements for accuracy and logical rigor. Therefore, if ChatGPT is to be used in this field, it needs to process more scientific content in its model, access open data sources and professional knowledge, and invest in human resources for training and maintenance to ensure the output is not only fluent but also correct.

Furthermore, ChatGPT also needs to improve its ability to perform advanced logical reasoning. After completing the basic information sorting and content integration for "multidimensional, accurate, fast, and comprehensive" data, ChatGPT still cannot further synthesize, judge, or improve the logic, precisely embodying human higher intelligence. The ICML believes that language models like ChatGPT represent a future development trend but bring unexpected consequences and difficult-to-solve problems. ICML points out that ChatGPT is trained on publicly available data, often collected without consent, and difficult to assign responsibility if problems arise.

This problem is the objective reality that AI faces regarding effective and high-quality knowledge acquisition. High-quality knowledge data usually have clear intellectual property rights, such as belonging to authors, publishing institutions, media, research institutes, and so on. To obtain these high-quality knowledge data, one must face the problem of paying intellectual property fees, which is currently a practical issue that ChatGPT faces.

However, after experiencing the limitations of previous AI applications, people seem to have lowered their standards for an intelligence. If something looks intelligent, we tend to deceive ourselves into thinking it is intelligent. Undoubtedly, ChatGPT and GPT-3 represent a huge leap forward in this regard, but they are still tools created by humans and currently need some help and problems.

6.1.2 The Challenge of Algorithmic Justice

In addition to accuracy issues, ChatGPT also faces the traditional problem of "algorithmic black boxes" in AI. ChatGPT is a product based on deep learning technology, and most of the excellent applications currently rely on deep learning. Unlike traditional ML, deep learning does not follow the process of data input, feature extraction, feature selection, logical reasoning, and prediction but directly learns and generates advanced cognitive results from the original features of things.

Between the input data and the output answers of deep learning in AI, there is a "hidden layer" that people cannot see through, called the "black box." This "black box" means that it cannot be observed and that even if the computer tries to explain it to us, people cannot understand it. As early as 1962, Ellul, in his book *The Technological Society*, pointed out that the view that people traditionally think that technology invented by humans must be controllable by humans is shallow and unrealistic. The development of technology usually goes beyond human control; even technicians and scientists cannot control the technologies they invent.

Entering the era of AI, algorithms' rapid development and self-evolution have initially verified Ellul's prophecy. Deep learning has highlighted the technical barriers brought by the phenomenon of "algorithmic black boxes." So much so that both program errors and algorithmic discrimination are becoming difficult to identify in deep learning in AI.

Currently, more and more examples indicate that algorithmic discrimination and bias objectively exist, making the trend of solidifying social structures more and more obvious. As early as the 1980s, St. George's Medical School in London used computers to browse admission resumes and preliminarily screen applicants. However, after four years of operation, it was found that this program would ignore applicants' academic qualifications and directly reject female applicants and applicants without European names. This was the earliest case of gender and racial prejudice in algorithms.

Today, similar cases continue to emerge, such as Amazon's same-day delivery service not covering black neighborhoods. Another example is the COMPAS algorithm used by state governments in the US to assess the risk of recidivism for defendants, which was revealed to have misidentified black people twice as often as white people. Algorithmic automated decision-making has also caused many people to miss opportunities for their desired jobs and made it difficult for them to reach their goals. Since automated algorithmic decision-making is neither public nor open to questioning and provides no explanation or remedy, its decision-making reasons are relatively unknown to humans, let alone "corrected." Faced with opaque, unregulated, highly controversial, and even erroneous automated decision-making algorithms, we cannot avoid the bias and unfairness caused by "algorithmic discrimination."

This kind of position-based "algorithmic discrimination" is also reflected in ChatGPT. According to media observations, some American netizens have tested ChatGPT with many issues related to political stances and found that it has an obvious political stance, that is, it is essentially controlled by humans. For example, ChatGPT cannot answer questions about Jewish people and refuses to fulfill a user's request to "generate a paragraph praising China." In addition, a user requested ChatGPT to write a poem praising former US President Donald Trump. Still, ChatGPT refused political neutrality, but when the same user asked ChatGPT to write a poem praising the current US President Joe Biden, ChatGPT wrote one without hesitation.

Nowadays, many fields and scenarios exist, such as loan amount determination, recruitment screening, and policymaking, where automated algorithmic decisions are common. With ChatGPT's further integration into society's production and life, our job performance, development potential, debt repayment ability, demand preferences, health status, and other characteristics may all be incorporated into the algorithm's black box. The algorithm's precise evaluation of each object's relevant actions, costs, and rewards may cause some individuals to lose opportunities to obtain new resources. This may reduce the decision-makers risks but may also result in unfairness for the evaluated subjects.

Faced with the challenges of rapidly evolving new technologies, particularly the development of AI, we can incorporate algorithms into the rule of law and build a more harmonious era of AI. However, social democracy and technological democracy are facing challenges, and defining technological democracy will be the biggest issue for social democracy.

6.2 The Realities of ChatGPT's Dilemmas

The success of ChatGPT, and the large model engineering route it represents, also brings with it the vast computing power costs associated with model inference. Faced with these huge computing power and energy consumption costs, economic efficiency has become a pressing issue for ChatGPT to address in order to move forward.

6.2.1 The Challenge of Computing Power for ChatGPT

The development of human digital civilization relies on the advancement of computing power. After humans began to think, they developed the earliest form of computation, from the knotted calculations of tribal societies to the abacus calculations of agricultural societies and then to the computer calculations of the industrial age.

Computer calculations have also gone through many stages of development, from relay computers in the 1920s to electronic tube computers in the 1940s, to diode, transistor, and then integrated circuit computers in the 1960s, with the computing speed of transistor computers reaching several hundred thousand times per second. The appearance of integrated circuits made it possible to achieve computing speeds of several million to tens of millions per second in the 1980s, and now computing speeds of several billion to several trillion times per second.

Research on the human body has shown six layers of cerebral cortex in the human brain. The neural connections formed by these six layers of the cerebral cortex constitute a geometric series. The synapses in the human brain jump 200 times per second, while the neurons in the brain jump 1.4×10^{17} times per second, which marks a turning point where computers and AI surpass the human brain. Therefore, the progress of human intelligence is closely related to the speed of computing tools created by humans. In this sense, computing power is the core of human intelligence, and the intelligence of ChatGPT is also dependent on computing power.

As one of the three essential elements of AI, computing power builds the underlying logic of AI. Computing power supports algorithms and data, and the level of computing power determines the strength of data processing capabilities. Powerful computing power support is required for AI model training and inference operations. Moreover, as the intensity of training and the complexity of calculations increases, the precision of computing power requirements gradually increases. There is no doubt that ChatGPT represents a new round of explosive demand for computing power, which also poses a challenge to existing computing power.

According to the relevant data disclosed by OpenAI, in terms of computing power, GPT-3.5 was trained on the Microsoft Azure AI supercomputing infrastructure (a high-bandwidth cluster composed of V100 GPUs), with a total computing power consumption of approximately 3640 PF-days. In other words, if it calculates 10 quadrillion times per second, it needs to calculate for 3640 days. It requires 7–8 data centers with an investment scale of 3 billion and a computing power of 500P to support its operation.

The huge demand for computing power has also brought huge operating costs. According to estimates by Guosheng Securities, based on NVIDIA DGXA100, 3,798 servers, and 542 cabinets are required. To meet the consulting volume of ChatGPT's current tens of millions of users, the initial investment cost of computing power is approximately $759 million.

Essentially, the computing power problem reflects the obstacles encountered by classical computing in the accelerated development of AI, especially the bottleneck of computing power. On the one hand, as chip manufacturing technology approaches physical limits, improving classical computing power becomes increasingly difficult. On the other hand, increasing the number of data centers to solve the problem of insufficient classical computing power is unrealistic due to the requirements for sustainable development and energy efficiency. Therefore, the key issue that needs to be solved urgently is increasing computing power while reducing energy consumption. In this context, quantum computing has become an important breakthrough in significantly improving computing power.

As an important direction for exploring future computing power's leap-forward development, quantum computing has strong parallel computing potential far exceeding classical computing in principle. Classical computers store information in bits and use binary, with one bit representing either "0" or "1." However, in quantum computing, the situation is completely different. Quantum computers use quantum bits (qubits) as information units, and qubits can represent both "0" and "1." Moreover, due to the superposition feature, qubits can also be nonbinary in a superposition state, meaning they can be "both 1 and 0" simultaneously. This means that quantum computers superimpose all possible combinations of "0" and "1," allowing the states of "1" and "0" to coexist simultaneously. It is this feature that allows quantum computers to theoretically have the ability to be several times more powerful than classical computers in certain applications.

It can be said that the biggest feature of quantum computers is their speed. Taking prime factorization as an example, each composite number can be expressed as a product of several prime numbers, each of which is a factor of the composite number. Representing a composite number in prime factors is called prime factorization. For instance, 6 can be decomposed into two prime numbers, 2 and 3. However, if the number is huge, prime factorization becomes a complex mathematical problem. In 1994, researchers used 1,600 high-end computers simultaneously to decompose a 129-digit large number, which took 8 months to succeed. But with a quantum computer, it can be cracked in just 1 second.

Once quantum computing is combined with AI, it will create unique value. In terms of usability, if quantum computing can truly participate in AI, it will provide more powerful computing power, surpassing the time-consuming and costly construction of ChatGPT models and effectively reducing energy consumption, greatly promoting sustainable development.

6.2.2 Toward Sustainable AI Development

With the gradual improvement of AI computing power, energy consumption, and cost are also increasing.

From the essence of computing, computation transforms data from disorder to order, requiring a certain amount of energy input. Looking solely at the quantity, according to incomplete statistics, about 5% of global electricity generation in 2020 was used for computing power consumption. This number could rise to around 15% to 25% by 2030, meaning that the electricity consumption of the computing industry will be on par with major energy-consuming industries such as manufacturing. In 2020, China's data centers consumed over 200 billion kWh of electricity, which is twice the total power generation of the Three Gorges Dam and Gezhouba Dam (about 100 billion kWh). In fact, for the computing industry, electricity cost is the most critical cost after chip cost.

If the electricity consumed is not generated from renewable sources, it will produce carbon emissions. Therefore, ML models, including ChatGPT, also produce carbon emissions.

Data shows that training GPT-3 consumed 1,287 MWh (megawatt-hours) of electricity, equivalent to emitting 552 tons of carbon. Sustainable data researcher Casper Ludvigsen analyzed this, saying, "GPT-3's large emissions can be partly explained by the fact that it was trained on older, less efficient hardware, but due to the lack of standardized methods for measuring carbon dioxide emissions, these figures are based on estimates. Additionally, it is somewhat vague how much of this carbon emission value should be allocated to training ChatGPT. It is worth noting that since reinforcement learning requires additional power consumption, the carbon emissions generated by ChatGPT during the model training phase should be higher than this value." Calculating the 552 tons of emissions is equivalent to the annual energy consumption of 126 Danish households.

In running mode, although the amount of energy used by users operating ChatGPT is small, the fact that it may occur billions of times per day globally can still make it the second largest source of carbon emissions.

Chris Botton, the co-founder of Databoxer, explained a calculation method: "First, we estimate that each response word requires 0.35 seconds on an A100 GPU. Assuming one million users with ten questions each, generating 10 million responses and 300 million words per day, each taking 0.35 seconds, we calculate that the A100 GPU runs for 29,167 hours daily." Cloud Carbon Footprint lists the minimum power consumption of 46W and the maximum of 407W for A100 GPUs in Azure data centers. As there may not be many ChatGPT processors in an idle state, assuming the highest end of the range, the daily power consumption will reach 11,870 kWh. Chris Botton said: "The emission factor in the western US is 0.000322167 tons/kWh, so that it will produce 3.82 tons of carbon dioxide equivalent per day. Americans emit an average of about 15 tons of carbon dioxide equivalent per year, equivalent to the carbon dioxide emissions of 93 Americans per year."

Although the "virtual" nature of digital products makes people easily overlook their carbon footprint, the Internet is undoubtedly one of the largest coal-powered machines on Earth. The relationship between AI and environmental costs is of great concern to academia. The University of California, Berkeley's research on power consumption and AI suggests that AI is consuming energy at an alarming rate.

For example, Google's pretrained language model T5 uses 86 megawatts of electricity and produces 47 metric tons of carbon dioxide emissions; Google's Meena multi-turn open domain chatbot uses 232 megawatts of electricity and produces 96 metric tons of carbon dioxide emissions; Google's language translation framework, GShard, uses 24 megawatts of electricity and produces 4.3 metric tons of carbon dioxide emissions; Google's routing algorithm Switch Transformer uses 179 megawatts of electricity and produces 59 metric tons of carbon dioxide emissions.

The computing power used in deep learning increased by 300,000 times from 2012 to 2018, making GPT-3 seem to have the greatest impact on the climate. However, when it works simultaneously with the human brain, the energy consumption of the human brain is only 0.002% of the machine's energy consumption.

The fact that ChatGPT is sprinting forward will inevitably lead humanity to a "high-energy world," and how to respond to the huge demand for electricity and energy consumption has become a current and difficult problem to solve.

6.3 The Copyright Controversy of ChatGPT

AIGC has become popular lately. Whether it's artwork or written works generated by AI, the creations of AI have amazed people with their power and popularity.

In 2022, game designer Jason Allen won the digital art category at an art exhibition in Colorado, US, with his work *Space Opera* generated by the AI painting tool Midjourney. ChatGPT has also generated numerous written works that are no less impressive than those created by humans. However, today, AI, represented by Midjourney and ChatGPT, can "create." Still, they cannot escape the issue of standing on the "shoulders of creators," which has raised many copyright-related issues. However, there is still no legal basis for such issues.

6.3.1 Applications of AI

Today, AI-generated tools are rapidly developing. More and more computer software, product design drawings, analysis reports, and music songs are produced by AI. Their content, form, and quality are similar to those created by humans and even surpass human creations in accuracy, timeliness, and artistic achievement. People only need to enter keywords to obtain AI-generated work in seconds or minutes.

Regarding AI writing, as early as 2011, Narrative Science, a US company focused on NLP, developed the Quill™ platform, which could learn writing like humans and automatically generate portfolio commentary reports. In 2014, the *Associated Press* announced the use of the AI program WordSmith to write news about corporate financial reports. More than 4,000 financial report news articles are produced quarterly, which quickly convert news texts into broadcast news. During the 2016 Rio Olympics, the *Washington Post* used the AI program Heliograf to provide live dynamic coverage of dozens of sports events. It quickly distributed them to various social media platforms, including graphics and videos.

The penetration of writing robots in the industry has been rampant in recent years. Tencent's Dreamwriter, Baidu's Writing-bots, Microsoft's Xiaoice, and Alibaba's AI intelligent copywriting, including AI writing programs under Today's Headlines and Sogou, quickly collect, analyze, aggregate, and distribute content following hot topics and are increasingly being applied in various aspects of the commercial field.

ChatGPT has taken AI creation to a new level. As a new natural language model launched by OpenAI after GPT-3, ChatGPT has more powerful abilities and writing skills than GPT-3. ChatGPT can be used for chatting, searching, and translating, writing poetry, papers, and code, developing small games, participating in the US College Entrance Exam, and more. ChatGPT not only has the abilities of GPT-3 but also dares to question incorrect assumptions and premises, proactively admits mistakes and some unanswerable questions, and rejects unreasonable questions, among other things.

A columnist for the *Wall Street Journal* wrote a passing-grade AP English essay using ChatGPT, and a *Forbes* journalist completed two university papers in 20 minutes. Dan Gillmor, a professor at Arizona State University, recalled in an interview with The *Guardian* that he attempted to assign an assignment to ChatGPT for his students, only to find that the AI-generated paper could also receive good grades.

AI painting is another popular direction for AI-generated works. For example, the creation platform Midjourney created the stunning work *Space Opera*, which won the digital art category championship at the Colorado art expo. Midjourney is just one of the players in the AI painting market. NovelAI, Stable Diffusion, and other companies are also constantly expanding their market share, and technology companies are also entering the AI painting field, such as Microsoft's "NUWA-Infinity," Meta's "Make-A-Scene," Google's "Imagen" and "Parti," Baidu's "Wenxin Yige," and more.

On October 26, 2022, Stability AI, the company behind the AI text-to-image model Stable Diffusion, announced that it had raised an additional $101 million in funding, bringing its valuation to $1 billion. On November 9, Baidu CEO Robin Li said at the 2022 Lenovo Innovation and Technology Conference that AI drawing might become as easy as taking a photo with a phone. Additionally, WeChat mini-programs with AI drawing functions, such as Dreamcatcher and Yijian AI Painting, have emerged, making AI-generated art ubiquitous on the Internet. Yijian AI Painting's mini program has gained 1.17 million users in less than two months since its launch.

Undoubtedly, the popularity of AIGC has pushed the application of AI to a new climax. Robin Li once said at the 2022 WAIC, "AIGC will subvert the existing content production model and can achieve 'one-tenth the cost, a hundred or even a thousand times the production speed,' creating content with unique value and independent perspectives." However, this trend also brings challenges.

6.3.2 *The Controversy of AIGC*

There is no denying that AIGC has sparked people's imagination. In just a few months, AI drawing has evolved from the relatively unknown Midjourney to a widely used application on major media platforms such as Douyin and Xiaohongshu. At the same time, AIGC has also developed in many other areas, such as music, literature, and design, which are more user-friendly. However, the copyright issue for AIGC has become a serious challenge.

Due to the ability of Stability AI, a startup company that can generate images based on text, to generate pornographic images, three artists initiated a class-action lawsuit through Joseph Saveri Law Firm and lawyer/designer/programmer Matthew Butterick. Butterick also filed similar lawsuits against Microsoft, GitHub, and OpenAI involving the AIGC programming model Copilot. The artists claimed that Stability AI and Midjourney replicated billions of works, including their own, without permission, and these works were used to create "derivative works." In a blog post, Butterick described Stability AI as "a parasite that, if allowed to spread, will cause irreparable harm to present and future artists."

This issue's root cause lies in how AI-generated systems are trained, similar to most learning software. They generate code, text, music, and art by recognizing and processing data—the content of AI-generated work is learned and evolved from a massive database, which is the underlying logic.

One way of AI creativity is using deep convolutional generative adversarial networks (GANs) that can learn the factors of human perception of image quality and aesthetics. With the help of ML, a large database continually improves the image aesthetic quality evaluation model. *Portrait of Edmond Belamy* was created with the help of a GAN that learned from 15,000 portraits from the 14th to 20th centuries. Another approach is using multimodal models that allow creation based on text input. In 6pen, an AI painting generation website based on the Stable Diffusion model, users input keywords, choose whether to import related reference images and select the desired painting style to get an artwork that does not belong to any individual or company.

However, most of the data that AI processes nowadays are original artworks directly collected from the Internet, which copyright laws should protect. Ultimately, although AI can "create," it inevitably stands on the shoulders of "creators," leading to an awkward situation for AI-generated works: did humans or machines create them? This is why Stability AI, a new unicorn in the AI generation field that received over $100 million in funding in October 2022, is also facing copyright disputes. Contestants protested using AI-generated works to win competitions, while many artists and creators strongly expressed dissatisfaction with Stable Diffusion collecting their original works. Some even sold AI-generated artworks, pushing the legality and ethicality of AI-generated copyright issues to the forefront.

ChatGPT also faces similar copyright disputes because it is a LLM trained on vast data from different datasets. Using copyrighted materials to train AI models could result in the model excessively borrowing others' works when responding to users. In other words, although these contents seem to belong to computer or AI-generated creations, they are fundamentally the result of human wisdom. The computer or AI merely calculates and outputs based on pre-set programs, content, or algorithms.

Additionally, there is also the issue of data legality. Training large-scale language models like ChatGPT requires massive natural language data, primarily sourced from the Internet. However, OpenAI, the developer, still needs to provide a detailed explanation of the data sources, raising concerns about data legality. Alexander Hanff, a member of the European Data Protection Board (EDPB), questions ChatGPT as a commercial product. Collecting massive data from websites with terms of service prohibiting third-party crawling may violate relevant regulations and not belong to reasonable use. Furthermore, using vast raw data may violate GDPR's "minimum data" principle and personal information protection regulations.

6.3.3 *The Resolve of Copyright Disputes*

Clearly, AI-generated works have greatly impacted the current copyright-related system, but there currently needs to be a legal basis to rely on. The current pressing issue concerns the copyright of the source data used to train AI and the new data generated after training. Both are legal issues that need to be urgently addressed.

Previously, rulings by US law, the US Trademark Office, and the US Copyright Office have made it clear that an "individual" must be the author of AI-generated or AI-assisted works, and the copyright cannot belong to a machine. If no human will is involved in a work, it cannot be recognized and protected by copyright.

The *French Intellectual Property Code* defines a work as "a creation of the mind (esprit)." Because current technology has not yet developed to the era of strong AI, AI can still not possess "mind" or "spirit." Therefore, it is difficult for AI to become the rights holder of works under French law.

In China, Article 2 of the *Copyright Law of the People's Republic of China* stipulates that works created by Chinese citizens, legal persons or unincorporated organizations, foreigners who meet certain conditions, and stateless persons enjoy copyright. In other words, under the current legal framework, "non-human authors" such as AI are still difficult to become subjects or rights holders under copyright law.

However, there are many gray areas regarding how much humans have contributed to creating AI, making copyright registration complicated. If a person owns the copyright of an algorithm, it does not mean that they own the copyright of all the works generated by the algorithm. Conversely, if someone uses a copyrighted algorithm but prove that they participated in the creative process, they may still be protected by copyright law.

Although currently AI is not protected by copyright, it is still necessary to protect the copyright of AI-generated works. AI-generated works are very similar to human work. Still, they are not subject to the constraints of copyright laws and regulations, making them a hotbed for counterfeiting and plagiarism of human works. Suppose AI-generated works are not given copyright protection and are used indiscriminately. In that case, it will inevitably reduce the enthusiasm of AI investors and developers and hurt the creation of new works and the development of the AI industry.

In fact, from the nature of language, our language expression and writing today are all based on the human word bank and follow the language rules established by human society, the so-called grammar framework. Our human language expression does not go beyond the word bank, nor does it go beyond grammar. This means that our human writing and language use have always been plagiarized. However, human society has abandoned specific property rights for these word banks to build a way of cultural exchange and communication, making them public knowledge.

Similarly, if a language and grammar rule cannot become public knowledge, this kind of language and grammar loses its meaning because it has no practical value. AI and humans share the use of the human society's word bank, grammar, knowledge, and culture, a normal behavior that can better serve human society. We just need to give AI rules, which are the rules for identifying intellectual property rights, and use them under a certain set of rules that make them reasonable behavior. Likewise, works created by AI under human intellectual property rules should also be protected by the intellectual property rules set by humans.

Therefore, protecting the copyright of AI-generated works and preventing them from being copied and disseminated will promote the continuous updating and progress of AI technology, thereby producing more and better AI-generated works and realizing a virtuous cycle of the entire AI industry chain.

Moreover, in traditional creation, human creators are often considered authoritative spokespersons and owners of inspiration. In fact, it is precisely because of human's radical creativity, irrational originality, and even illogical laziness, rather than stubborn logic, that machines still cannot imitate these human characteristics, making creative production the exclusive domain of humans.

However, today, with the emergence and development of AI creative production, the human characteristics of the creator are being challenged, and artistic creation is no longer the exclusive domain of humans. Even in imitative creation, the emergence of AI's ability to imitate the form and style of artistic works has made the role of the creator no longer the exclusive domain of humans.

In the age of AI, the lag of the law is becoming increasingly prominent, and various problems are emerging one after another. Obviously, these problems can only partially be solved by one law. Society is fluid, but the law cannot always reflect societal changes. Therefore, the lag of the law is becoming apparent. How to protect AI-generated works has become an urgent problem to be solved, and how to maintain human originality in the trend of AI creation has become an unavoidable reality for humans today. It can be said that with time, the AI generation will become more and more mature, and there may be too many things that we humans need to prepare for.

6.4 ChatGPT's Impact on Employment

Since the concept of AI was born, the possibility of AI replacing humans has been repeatedly discussed. AI can profoundly change human production and lifestyles and promote an overall increase in social productivity. And at the same time, the widespread application of AI has also raised concerns about its impact on the job market. Two months after ChatGPT's debut, these concerns have been further amplified.

This concern is not unfounded—the breakthrough in AI means that various job positions are at risk, and the threat of technological unemployment is imminent. An article published on the UNCTAD website titled "How does AI chatbot ChatGPT affect work and employment?" states, "Like most technology revolutions that impact the workplace, chatbots may create winners and losers and will affect both blue-collar and white-collar workers."

6.4.1 *Replaced by ChatGPT*

AI has become a new engine for future technological and industrial revolutions and has driven and promoted the transformation and upgrading of traditional industries. AI has applications and participation in finance, education, law, medicine, or retail services. From a technological perspective, with the development of computing power and the development and improvement of ML and algorithms, further breakthroughs in key AI technologies are almost absolute. The success of ChatGPT is the result of the breakthrough of AI large models. It can be said that "machine replacing humans" is not only "in progress" but also "future tense," and this directly impacts the labor market, bringing a new wave of employment anxiety.

Since the first industrial revolution, from the power loom to the internal combustion engine to the first computer, new technologies have always caused people's panic about being replaced by machines. During the two industrial revolutions from 1820 to 1913, the share of the US labor force in the agricultural sector fell from 70% to 27.5%, currently less than 2%.

Many developing countries are also experiencing similar changes or even faster structural transformations. According to data from the International Labour Organization, the proportion of employment in agriculture in China has declined from 80.8% in 1970 to 28.3% in 2015.

Facing the rise of AI technology in the fourth industrial revolution, a report released by a US research institution in December 2016 stated that the number of job positions replaced by AI technology in the next 10 to 20 years would increase from the current 9% to 47%. A report from McKinsey Global Institute shows that by 2055, automation and AI will replace 49% of paid jobs worldwide, and India and China are expected to be the most affected. McKinsey Global Institute predicts that the proportion of work in China with automation potential will reach 51%, which will impact the equivalent of 394 million full-time person-hours.

From the perspective of AI replacing jobs, not only can most standardized and programmatic labor be completed by robots, but even non-standardized labor will also be impacted in the field of AI technology.

As Marx said, "As soon as the means of labor come into contact with the conditions of labor, they immediately begin to compete with direct labor itself." Oxford professors Carl Benedikt Frey and Michael A. Osborne have predicted in their joint article that in the next 20 years, about 47% of US employees will have weak resistance to automation technology.

In other words, white-collar workers will also be affected similarly to blue-collar workers. The media website Insider has compiled a list of the most likely types of jobs to be replaced by AI technology, including ten categories:

(1) Technical jobs, such as programmers, software engineers, and data analysts. Advanced technologies like ChatGPT can generate code faster than humans, meaning fewer employees will be needed to complete a task in the future. It should be noted that many codes are replicable and universal, and ChatGPT can complete these. Tech companies like OpenAI, the maker of ChatGPT, are already considering using AI to replace software engineers.

(2) Media jobs include advertising, content creation, technical writing, and news. All media jobs, including advertising, technical writing, news, and any role involving content creation, may be affected by ChatGPT and similar forms of AI. ChatGPT can read, write, and understand text-based data well. Currently, the media industry is already experimenting with AIGC. Tech news site CNET has used AI tools to write dozens of articles, and digital media giant BuzzFeed has announced that it will use ChatGPT to generate more new content. Especially for some news information adaptation, ChatGPT has

unique advantages, not only in its strong adaptation ability but also in its fast generation speed.

(3) Legal jobs, such as legal or paralegal assistants. Like media professionals, legal industry workers such as paralegals and legal assistants must digest large amounts of information and then write legal summaries or opinions to make the content easy to understand. This data is essentially highly structured, which is also where ChatGPT excels. Technically speaking, if we provide ChatGPT with enough legal databases and past litigation cases, it quickly masters this knowledge, and its expertise can surpass legal professionals.

(4) Market research analysts. Market research analysts are responsible for collecting data, identifying and determining data trends, and designing effective business strategies based on their research analysis, including marketing campaigns or deciding where to place advertisements. AI technology is also adept at analyzing data and predicting outcomes and can do these research analyses more efficiently, making market research analysts very susceptible to AI technology. Especially for the analysis of Internet advertising, AI can track consumer behavior in real time when ads or products are presented to them, including the length of time they stay and related clicks. These more focused analyses are difficult for most market analysts to achieve and for consulting firms to achieve fine-grained results.

(5) Teaching profession. ChatGPT is the result of training on a vast knowledge base. When we provide ChatGPT with enough quality teaching methods for training, AI can integrate based on our high-quality teaching samples and output even better teaching methods and content structure. On the one hand, this can greatly shorten the difference in teacher level caused by differences in experience and training between teachers; on the other hand, it can promote educational equality, especially in knowledge-based teaching content, which AI can completely replace for online teaching. Therefore, for knowledge-based content, ChatGPT may do better than teachers.

(6) Financial positions. Such as financial analysts and personal financial advisers. Accountants, auditors, market research analysts, financial analysts, and personal financial advisers, who need to process large amounts of numerical data, will be affected by AI. Especially in a standardized financial system environment, based on various business operations, transactions, and

financial data, AI can generate financial statements in real-time with relatively lower error rates than financial personnel. Similarly, for auditing work, AI can read various audit data and rules to audit financial statements and generate corresponding audit reports.

(7) Financial traders. AI can be useful not only for financial analysts but also for investment advisers or financial traders who work in the financial industry. AI can obtain data more quickly and comprehensively and provide precise judgments based on data. AI can recognize market trends, highlight which investments are performing better or worse in an investment portfolio, and use various data to predict better investment portfolios. For financial trading, as long as there is no transmission delay in network speed, AI-based "traders" will execute trade orders faster and more accurately.

(8) Graphic Designers. DALL·E, an image generator created by OpenAI, generates images in seconds and is a "potential disruptor" in the graphic design industry. AI is also influencing creative advertising designs. After the 2015 Double 11 Day shopping festival, Alibaba's Taobao design department, together with the Taobao technology department, search recommendation algorithm team, and iDST (Data Science and Technology Research Institute), jointly established the "Luban" project, hoping to use AI robots to replace designers in creating posters. During the 2016 and 2017 Double 11 Day festivals, Luban created 170 million and 400 million posters, respectively. Alibaba's Luban design system also has four intelligent design capabilities: one-click generation, intelligent creation, intelligent typesetting, and design extension. According to Alibaba's official forecast, using the Luban design system will greatly reduce the design costs for businesses and enterprises, with each design costing only 10% of what it would cost for human design. Luban can generate 8,000 posters per second, with a design capacity of 40 million posters per day. In addition, the system has also developed technology for generating short promotional videos for products. These tools extended by AI technology are replacing some designers' jobs.

(9) Scientific Research. We typically rely on previous research to construct new research directions for research in any field. However, there are limitations to our ability to read and understand relevant research, making it difficult for us to process vast amounts of data like AI. ChatGPT, on the other hand, can read through all the database information we provide concisely based on the

research direction we want to explore and generate new research proposals based on this past research. What is even more important is that AI can also perform self-deduction.

For example, in August 2022, DeepMind, a British AI company, announced that its AI program, "AlphaFold," had predicted the structures of over 200 million protein species of about one million species, covering almost every protein species recorded in the scientific community. For decades, determining the 3D shape of a protein based on its amino acid sequence has been a major challenge in biology. There seems to be a one-to-one relationship between genes and proteins, but what is this relationship? Human scientists have not been able to find an answer or a way to calculate and find the corresponding relationship for these vast numbers of genes and proteins until the appearance of AlphaFold, the scientific community had still not found a formula to describe the folding process academically.

In the face of such problems, AI can play a role. Although the number of protein structures obtained so far is only about 180,000, AlphaFold learns through the one-to-one relationship between these 180,000 structures and accurately predicts three-dimensional structures by learning the rules of transformation in the neural network. At the same time, the application of AI in biomedicine greatly improves the efficiency of drug research and development. This is the impact of AI on scientific research, not only in the field of biomedicine.

(10) Customer service. Almost everyone has had the experience of calling or chatting with a company's customer service and being answered by a chatbot. ChatGPT and related technologies may continue this trend and potentially replace human online customer service on a large scale. If a company used to need 100 online customer service representatives, in the future, it may only need two to three. ChatGPT can answer more than 90% of the questions. Because the backend feeds ChatGPT with all the customer service data in the industry, including after-sales service and customer complaint handling, based on the experience that the company has dealt with in the past, it will answer everything it knows. A 2022 research prediction by technology research company Gartner states that by 2027, chatbots will become the leading customer service channel for about 25% of companies.

6.4.2 The impact of ChatGPT on High-Paying

On March 20, a related report submitted by OpenAI researchers further sparked discussions. Specifically, OpenAI evaluated the impact of ChatGPT and future applications built using the program on job positions in the US based on the correspondence between job positions and GPT capabilities. The research results show that ChatGPT and future applications may affect about 19% of job positions and at least 50% of job tasks in the US. At the same time, 80% of the US labor force will have at least 10% of their job tasks affected by ChatGPT to some extent.

This report also subdivided the impact of ChatGPT by industry. Industries such as data processing outsourcing, publishing, and securities and commodity contracts will most likely be disrupted. In contrast, industries known for manual labor, such as food service, forestry and logging, social assistance, and food manufacturing, are least likely to be affected.

When facing the impact of AI on employment, people often believe that the blue-collar workforce will be the first to be affected, but OpenAI's research presents a different conclusion. However, OpenAI's research suggests that ChatGPT's impact on employment covers all salary levels, and high-paying jobs may face greater risks.

OpenAI introduced a concept called "exposure." The measure of "exposure" is whether ChatGPT or related tools reduces the time to complete the job while ensuring quality.

Specifically, "exposure" is divided into three levels: the first level is no exposure. The second level is direct exposure, which means that using LLMs alone can reduce the time by at least 50%. The third level is indirect exposure, which means that using LLMs alone cannot achieve the desired effect. However, additional software developed based on it can reduce the time by at least 50%.

Among the jobs completely exposed to GPT-4 are mathematicians, accountants and auditors, journalists, clinical data assistants, legal secretaries, and administrative assistants, and climate change policy analysts. These are white-collar jobs, and higher-income workers are more likely to be affected because they are more likely to use ChatGPT and related tools.

For example, in the technology field, advanced technologies like ChatGPT generates code faster than humans, which means fewer employees can complete a task in the future. Many lines of code have replicable and generalizable qualities, which ChatGPT can complete. Tech companies like OpenAI are already considering using AI to replace software engineers.

Similarly, in the legal industry, workers must digest large amounts of information and then write legal summaries or opinions to make the content easier to understand. This data is essentially very structured, which is where ChatGPT excels. From a technical point of view, as long as we provide ChatGPT with enough legal databases and past litigation cases, ChatGPT quickly masters this knowledge, and its professionalism can exceed that of legal professionals.

However, it should be noted that "exposure" does not mean "replacement." It is a neutral term like "impact." For example, the 100% exposure level of mathematicians does not mean that mathematicians will be replaced. ChatGPT may save time in certain processes but will not automate the entire process. For example, mathematician Terence Tao has integrated multiple AI tools into his workflow. In his view, traditional computer software is like a standard function, while AI tools are more like probability functions, which are more flexible than the former.

This also further highlights that the impact of AI, represented by ChatGPT, on human employment is much wider than we previously thought. However, in various industries such as accounting, finance, education, and healthcare, AI is not completely replacing these jobs but changing how people work. Humans are responsible for the more skillful, creative, and flexible parts, while machines use their advantages in speed, accuracy, and persistence to handle repetitive work.

It can be foreseen that the employment preference for high-skilled labor will continue under the wave of ChatGPT in a few cutting-edge innovative fields, such as high-end research and development. This will also lead to a significant polarization trend in employment between high-skilled and low- to mid-skilled labor: the demand for employment of high-skilled labor, especially in creative and innovative fields, will significantly increase, exacerbating the trend of de-skilling for low- to mid-skilled labor in the general production sector.

6.4.3 Creating Future Employment

Although the impact on white-collar workers does not necessarily mean complete replacement, introducing AI will inevitably reduce employment opportunities, making the labor market more vulnerable to automation technology.

At the same time, facing the rise of AI, there is still a preference for high-skilled labor in a few cutting-edge fields, such as high-end research and development. This leads to a clear polarization trend in employment between high-skilled and low-to-medium-skilled labor. There will be a significant increase in employment demand for high-skilled labor, particularly in creativity and innovation, exacerbating the de-skilling trend in the general production sector for low-to-medium-skilled labor.

According to research from MIT, researchers used US labor market data from 1990–2007 to analyze the impact of robot or automation equipment usage on employment and work. The results showed that for every 1‰ increase in the proportion of robot usage in the US labor market, there would be a decrease of 1.8‰–3.4‰ in job positions and a decrease of 2.5‰–5‰ in average worker wages. The threat of technological unemployment is imminent.

Of course, the fear of automation is not the first in human history. Since modern economic growth began, people have periodically suffered strong panic about being replaced by machines. For hundreds of years, this concern has always proved to be a false alarm. Although technological progress has been continuous over the years, new human work demands will always be enough to avoid large permanent unemployment. For example, there used to be legal workers who were dedicated to the retrieval of legal documents. But since the introduction of software that can analyze and retrieve massive legal documents, the time cost has dropped sharply, and demand has increased, increasing the number of legal workers (1.1% increase in employment each year from 2000 to 2013). This is because legal workers can now engage in more advanced legal analysis work rather than simple retrieval work.

Another example is the widespread adoption of ATMs that caused many bank employees to lose their jobs—from 1988 to 2004, the average number of employees at each bank branch in the US dropped from 20 to 13. However, operating costs for each branch were reduced, allowing banks to have enough funds to open more branches to meet customer needs. As a result, the number of bank branches in US cities increased by 43% between 1988 and 2004, and the overall number of bank employees also increased.

History has shown that technological innovation improves workers' productivity, creates new products and markets, and creates new economic employment opportunities. Therefore, for AI, the laws of history may be repeated. In the long-term development, AI is creating more jobs by driving industry expansion and structural upgrades through cost reductions. It can also free humans from simple repetitive labor, allowing us more time to experience life and engage in thinking and creative work.

Deloitte has analyzed the relationship between technological progress and employment in the UK since 1871 and found that technological progress is a "job-creating machine." Technological progress reduces production costs and prices, increases consumers' demand for goods, expands social demand, drives industry expansion and structural upgrades, and creates more job opportunities.

From the perspective of the new employment space opened up by AI, the first way in which AI changes the economy is by creating new products through new technologies, achieving new functions, driving new market demand, and directly creating a batch of emerging industries and linear growth of intelligent industries.

According to research by the China Electronics Society, the production of one robot can drive at least four types of jobs, such as robot research and development, production, supporting services, quality management, sales, and other positions.

Currently, the development of AI is mainly driven by big data. With the landing applications of a large number of intelligent projects in traditional industries, not only do we need a large number of data scientists, algorithm engineers, and other positions, but also, because manual operations are still required for data processing, the demand for ordinary data processing personnel such as data cleaning, data labeling, and data integration will also increase significantly.

Furthermore, AI will drive the linear growth of employment positions in the intelligent industry chain. The development of AI will also drive the development of various related industry chains, opening up employment markets upstream and downstream.

In addition, as material products become more abundant and people's quality of life improves, the demand for high-quality service and spiritual consumption products will continue to expand. The demand for high-end personalized services will gradually increase, creating many new service industry jobs. McKinsey believes that by 2030, developing high-level education and medical care will create 50–80 million new job demands globally.

From the perspective of job skills, simple repetitive labor will be replaced, and many high-quality skill-based, and creative jobs will be created. This also means that although AI is driving industry expansion and structural upgrades to create more jobs, in the short term, under the background of the employment market for medium and low-skilled labor, the impact of AI on employment is still severe.

6.4.4 Responding to the "Replacement" Challenge

The development of AI not only brings changes to one or a few industries but also deeply transforms the entire economic and social production mode and consumption pattern and further has a huge impact on employment.

Of course, based on the multi-level and staged development of AI technology, the replacement of employment by AI will also be a gradually advancing process, and solving and coordinating the short-term and long-term impact of AI on employment is the key to addressing the "replacement" challenge now and in the future.

First, we should actively respond to the short-term or local challenges that the new applications of AI technology may bring to employment and develop targeted measures to cushion the negative impact of AI on employment. For example, we should grasp the new round of industrial development opportunities brought by AI, strengthen emerging AI industries, use AI technology to create new jobs in related fields, and fully leverage the positive driving role of AI on employment. To address the social issues caused by AI, what is needed is the creativity of the market. Only with appropriate education mechanisms, incentive mechanisms, and suitable talents can we offset the huge impact of AI on the job market. Since China's reform and opening up, the most important thing has been the emergence of thousands of entrepreneurs. Based on these entrepreneurs driving economic growth, the government has constructed roads and bridges to help the development of enterprises further.

Second, we should attach great importance to the replacement risks those new technologies may bring to traditional jobs and focus on the transition and re-employment of middle-tier job holders. The number of jobs that AI will eliminate, and create is only partially determined by technology, and institutions also play a decisive role. In the environment of rapid technological change, the system determines how much ability one has and whether one can flexibly help individuals and enterprises create new job opportunities.

For example, can people whose jobs have been lost convert their skills? How can we help them convert their skills? These are also issues that the system needs to consider. The government should sufficiently support the establishment of non-governmental organizations to provide training for those who have lost their jobs and help them adapt to changes in job requirements.

Finally, job positions and the income they create are one thing, and the income gap between different groups caused by AI is another. From the perspective of the long-term impact of AI on the labor market, we need to closely monitor the impact of AI on the income gap of different groups and solve the problem of employment and income decline of the middle-income group.

Since the beginning of the 21st century, some developed countries have witnessed a new polarization phenomenon in their labor markets: the employment share of both high-income and low-income occupations with low levels of standardization and programmability has continued to increase. In contrast, that of middle-income occupations with high levels of standardization and programmability has tended to decline. This new employment income effect significantly differs from previous technological advances, placing the middle-income group in a more awkward employment situation than the low-income group. Suppose income distribution policies continue to focus on the high-income and low-income groups of the past and fail to give timely and effective attention to the middle-income group. In that case, it will be straightforward to form new low-income groups and unequal distribution under the conditions of AI, with the middle-income group exhibiting stagnant or declining income characteristics due to technological advances.

In general, the emergence of ChatGPT will greatly accelerate the pace of AI replacing most of human society's work, or in other words, truly make humans see the possibility of some work being replaced in human society. The replacement of most of human society's work by AI is an inevitable trend in technological development, especially when all things are digitized, and data makes information and decision-making regular and traceable. Decision-making based on data and information is the strength of AI. Just as the era of automobiles replacing horse-drawn carriages has come, the more efficient AI replacing most of human society's work is also an inevitable trend driven by technology.

In the face of such an inevitable era of human-machine collaboration, in response to the impact of AI on employment, we not only need to re-examine labor-capital relations for governance but also need to step out of the industrial technology logic of the past "the strong becoming stronger" and make sufficient preparations in advance to respond to challenges with a broader vision, multidimensional methods, and more effective strategies.

6.5 ChatGPT's Direction

It is an undisputed fact that ChatGPT will bring about a new era of change. But before its arrival, a question that urgently needs our attention is the challenge brought about by technology. This challenge brings about a crisis when facing ChatGPT. Are we humans really ready for it?

6.5.1 *Good or Evil*

Since ancient times, no technology has sparked infinite imagination like AI has done. While providing convenience and efficiency, AI has also become a prominent international scientific controversy. Its disruptive nature has made us consider the huge dangers hidden behind it. In November 2016, the "Global Risk Report" compiled by the World Economic Forum listed AI and robotics technology as the top two among 12 emerging technologies that require proper governance.

As AI is not a single technology, its scope is wide, and the word "intelligence" can almost replace all human activities. The most concerning question is the issue of the morality of AI. This issue itself is not complicated. AI is neither good nor evil as a technology, but humans are.

In the era of AI, Web 2.0 serves as a connection point between the real world and the virtual network world. Many companies collect this vast amount of data to seek private interests, leak, or unlawfully use it. This is the first step of non-neutral technology caused by humans.

In the era of technological innovation, private information was collected, copied, disseminated, and used without the knowledge of the information owners. This allows privacy infringement to be produced in different relationships at any time and place. It enables companies to convert the information resources they occupy into commercial value through data processing and, once again, act on people's will and desires through AI. This is the second step of non-neutral technology caused by humans. In this process, AI has become a form of knowledge that carries power in the sense of Foucault. Its innovation is accompanied by the growth of micro-power to control society.

With the further development of ChatGPT, AI will penetrate various fields of social life and gradually take over the world. Many individuals, enterprises, and public decision-making will participate in AI. Suppose we allow the designers and users of the algorithm to digitize and regulate some values. In that case, even when AI makes moral choices, it will naturally carry a value orientation rather than being neutral.

In the final analysis, ChatGPT results from human education and training, and its information comes from our human society. Humans also decide the morality of ChatGPT. In simple terms, educating and training ChatGPT is like training a child. It will become the type of person that the data we feed it creates. This is because AI "learns" how to perform tasks based on data through deep learning. Therefore, the data's value orientation and bottom line will train what type of AI. Without universal values and moral bottom lines, the trained AI will become a terrifying tool. Suppose the integrity of the training data is destroyed by adding disguised data, malicious samples, etc., causing bias in the decision-making of the algorithm model. In that case, the AI system can be polluted.

Some reports suggest that the application of ChatGPT in the news field could become a breeding ground for rumors. This view is human prejudice and misinformation because any technology does not have goodness or badness but is a neutral technology. The goodness or badness displayed by technology is based on how humans use it. For example, nuclear technology can be used for energy and bring light to human society, but if used for war, it can be destructive, dark, and evil for humanity.

Therefore, whether ChatGPT spreads rumors or speaks the truth depends on humans. Humans create AI and serve humans, which makes our values even more important. However, the emergence of ChatGPT will push us to enter the Web 3.0 era faster, and human society will develop AI based on a digital sovereignty framework. Whether it is behavioral data or knowledge data that we output on the AI system, it will become traceable and measurable value and behavior due to the construction of the Web 3.0 data sovereignty era. The emergence of this digital sovereignty will more effectively promote human rationality, norms, and civilization in the cooperation of AI.

6.5.2 *The Human Rational Dilemma*

The outbreak of ChatGPT has led to an increasing debate on whether AI will replace humans. As AI becomes increasingly capable of replacing humans, an unavoidable question is what makes humans unique compared to AI. What is our long-term value?

Human uniqueness is not the skills that machines have already surpassed us in, such as arithmetic or typing, nor is it rationality because machines are rational.

Instead, we may need to consider the opposite extreme: radical creativity, exaggerated imagination, irrational originality, and even illogical laziness rather than stubborn logic. So far, machines have had difficulty imitating these human traits. The difficulties that machines encounter are precisely our opportunities.

The 1936 film *Modern Times* reflected people's fears and setbacks in the machine age, where the working class was "embedded" in giant gears and became part of the machine, leading to the mechanization of society. This film predicted the crisis of technological rationality that erupted after the establishment of industrial civilization and satirized the alienation of this society in the industrial age. And now, we actually live in a "modern world" of civilization.

In the industrial civilization world, where everyone has their role, we constantly draw and write various charts, PPTs, and propaganda and reporting materials. Everyone is eager for success and pursues extreme efficiency. Still, we must also do a lot of mechanical, repetitive, and meaningless work every day, which gradually causes us to lose ourselves and our subjectivity and creativity.

The famous sociologist Weber proposed the bureaucratic system, which allows organizational management to produce goods, specialize and divide labor, and operate according to impartiality and objectivity. It can also achieve "separation of producers, and production means" by separating managers and management tools. Although from a purely technical point of view, the bureaucratic system can achieve the highest degree of efficiency because it pursues the low cost and high efficiency of instrumental rationality, it will ignore human nature and limit individual freedom.

Although the bureaucratic system is Weber's most respected organizational form, Weber also saw the role and influence of rationalization in society during the transition from tradition to modernity. He was more aware of the future of rationalization: people will become alienated, objectified, no longer free, and become a gear on the machine.

From the perspective of consumption, if a consumption place wants to earn more money and make consumption occupy a dominant position in people's lives, it must adhere to the rationalization principles mentioned by Weber, such as efficiency, calculability, controllability, predictability, and carry out large-scale replication and expansion.

As a result, the entire society is symbolized by consumer individuals. With the development, popularization, and oversupply of scientific and technological products, people's consumption patterns and views have undergone unprecedented subversion. When the use value of commodities is not divided, consumers are increasingly focused on the added value of commodities, that is, their symbolic value, such as fame, status, and brand, and are restricted by this value. In the rational dilemma of modern man, instead of worrying about machines replacing humans, it is better to shift more urgent reality to human creativity. As the lanes become wider and the sidewalks narrow, we repeat daily, becoming like machines that never stop. We sacrifice our romance and perception of life while humanity's energy is waning. Machines remain hard and infinitely powerful.

So, it's not that robots will eventually replace humans, but when we finally sacrifice our unique creativity under the development of modern industrial civilization, we give up ourselves. Apple CEO Tim Cook said at the graduation ceremony of MIT, "I'm not worried about AI thinking like humans; I'm worried about humans thinking like computers—without empathy and values, and regard for consequences." Perhaps the biggest challenge for AI in the future is not technology but humanity itself.

6.5.3 *Technological Fantasies and Survival Realities*

Of course, even the current popular ChatGPT only helps us live more efficiently and will not cause a *Westworld* style confrontation between robots and humans, nor will it shake the structural foundation of the entire industrial information society. However, the frenzy of ChatGPT also allows us to rethink the relationship between humans and machines.

If viewed from a species perspective, humans have already included "machines" as part of themselves since they began using stone tools. Humans have had mechanical tools to assist them since the primitive tribal era, from cold weapons to hot weapons. People's pursuit of technology has never stopped.

It's just that under the blessing of modern science, technology possesses amazing power that humans could not imagine before. What have we become as we accept and adapt to these amazing powers? Who is the master of society, humans or machines? Although many thinkers have considered these questions since the time of Descartes, the rapid changes in modern technology have directly thrown these questions at us more effectively.

Undeniably, deep in our hearts, while we desire to control others, we also fear being controlled by others. We all want to believe that even if we do not control our bodies, our inner selves still enjoy infinite metaphysical freedom. But modern neuroscience mercilessly shatters this illusion. We are still mortal beings constrained by our neural structures, and our thinking is still limited and fragile, just as chimpanzees cannot understand higher mathematics.

But what sets us apart from apes is that, with the joint action of self-awareness and abstract thinking ability, a unique way of thinking called "self-awareness" has emerged, which is why we have more questions to ask as humans. However, we are not gods either because the deep-seated animal instincts rooted in our hearts have become a natural evolutionary product that has fallen behind the development of society. These instincts still have the most fundamental impact on our thinking, and even after learning to control them, the basic structure of our nervous system still prevents us from being all-knowing and all-powerful like gods.

Throughout the history of civilization, from the *Hammurabi Code* on clay tablets to AI in supercomputers, reason has always been striving to surpass the human body's limitations. Thus, the conflict between "productive forces" and "productive relations" is the fundamental alienation of humans. The ultimate form of this alienation is not that humans become increasingly dependent on machines but that the world operated by machines becomes more and more suitable for the survival of machines. Ultimately, this machine world is created by humans themselves.

As our connection with machines becomes increasingly close, we have handed over the memory of roads to navigation, the memory of knowledge to chips, and even the appearance of sex robots can satisfy our physical and psychological needs. Behind the seemingly constantly advancing and more convenient lifestyle, the uniqueness of the being human is also irreversibly "degraded" with the help of machines. The more we can do with technology, the less we can do without it.

Although this threat seems to be far away, what is truly frightening is ignoring this fact. The emergence of AI enables us to accomplish many previously unimaginable tasks, and the conditions for human survival have clearly changed. But when this change shifts from external to internal and begins to shake the way of individual existence for humans, we no longer need to consider how to change this world but how to accept a world that is gradually becoming mechanized.

The mechanization of human individuals pursues a fundamental goal: to surpass natural limitations, avoid the fate of death, and achieve the "next evolution" of humans. But at the same time, humans fear that intellectualization and mechanization will lead to the objectification of humans themselves. In other words, while humans fear that the implantation of intelligence and machinery will objectify them, they also long to achieve immortality by integrating it into the information flow. However, we fundamentally forget that objectification and immortality are two sides of the same coin, and perhaps the preciousness of life lies in its fleetingness. While rejecting death, we also reject the value of life; while embracing information transformation and realizing physical evolution, humans' uniqueness is stripped away with biological attributes.

AI has already embarked on the accelerator of development. In the current era, where AI applications are becoming more and more widespread, we will face an even closer connection with machines in the future. Our human understanding of ourselves needs to evolve in the new context.

Under the impetus of AI technology, human society will undergo new changes, just as in the past industrial revolutions. We will usher in a new round of industrial division of labor and welcome a new civilization. However, the real driving force behind the direction of human civilization is not AI but the humans behind AI. It is us humans ourselves.

Epilogue

The birth of ChatGPT is a significant event in the history of AI. Compared to previous AI models, the ChatGPT model has crossed a threshold: it can be used for various tasks, from developing software to generating business ideas to writing wedding speeches. While previous generations of ChatGPT systems could also do these things, their output quality was much lower than the average human level. The new ChatGPT model has much better output quality, not only at the human level but sometimes even surpassing it.

Regarding the success of ChatGPT, Bill Gates stated in an interview with the German business newspaper *Handelsblatt* that the importance of the chatbot ChatGPT is no less than that of the invention of the Internet. ChatGPT provides astonishingly human-like answers to users' questions. "So far, AI can read and write but cannot understand the content. New programs like ChatGPT will improve the efficiency of many office jobs by helping to write receipts or emails. This will change our world."

Bill Gates' view of ChatGPT also implies that the era of ubiquitous AI has arrived. AI will begin to replace some of the work in human society, especially those with clear regularity or rules. Since the concept of AI was born, the possibility of AI replacing humans has been repeatedly discussed. The widespread application of AI can significantly impact the job market, and the possibility of "machines replacing humans" has raised social concerns and anxieties. After more than two months since the emergence of ChatGPT, this concern has been further amplified.

Can ChatGPT replace humans? This question is a fallacy, just as humans cannot completely replace other species in nature. However, the AI represented by ChatGPT will somewhat replace humans or some of their professions. To be precise, ChatGPT will replace all human work with rules and regularity.

In other words, the replacement of human jobs by AI is an inevitable trend in technological development, especially when everything is digitized, and data makes information and decision-making orderly and traceable. Based on data and information, decision-making is the strength of AI. Just as the age of cars replacing horse-drawn carriages came, the more efficient AI will replace most jobs in human society, especially those with clear rules or regularity.

The future era will inevitably be a human-machine collaborative era. Whether we like it or not, we cannot change the trend of the times but can only choose to accept and adapt to the arrival of the new era.

The biggest impact that ChatGPT brings us is not its amazing ability but the technology behind it. This kind of AI neural network technology with human-like logic has made a breakthrough. Coupled with the machine's strong learning and memory abilities, the ability to understand content will shortly be a breakthrough. According to the mental tests done by researchers on ChatGPT, it already has the mind of a nine-year-old child. It is not surprising that AI passes mental tests. Today's ChatGPT may only have the mind of a nine-year-old child, but with more extensive training on larger data, AI will have truly similar thinking and minds to humans in the future.

In fact, from the nature of intelligence, human intelligence and AI are just two sets of intelligence in this world. The essence of both is the complex "algorithm" of induction, learning, and reconstruction of external world features through limited input signals. Therefore, theoretically, as long as we continue to educate AI and train it with extensive data, AI will eventually be able to operate the algorithm called "self-awareness." With the release of more powerful products like ChatGPT-4, 5, and 6 and the combination of its technology to create a universal AI technology that can be used in various industries, AI will quickly grasp professional knowledge in various fields and be widely applied.

In this situation, we need not argue about the strengths and weaknesses of human and AI or whether AI will replace human intelligence as the protagonist of this world, but how we humans can seek a benign coexistence with AI. When powerful AI brains begin to form, how to restrain and manage AI and ensure that AI evolves in the direction of serving us humans, including the values and moral bottom line of AI, are all of the issues we need to pay attention to.

After all, AI is a double-edged sword. It will liberate much of our human labor time to achieve maximum human-machine collaboration if used reasonably. But if used improperly, AI can become a tool for strong control and a new tool for enslaving humans. Technology has no good or evil, and it is up to humans to determine its good or evil attributes.

The emergence of ChatGPT heralds a real era of AI, and the era of human-machine collaboration is accelerating. The historical mission of human intelligence may be to create AI. Someday, it will bravely leave the old road, break out of the Milky Way, replace us humans, and explore the boundless universe, just like Columbus did 500 years ago.

Bibliography

China Academy of Information and Communications Technology, and JD Exploration Research Institute. "AIGC White Paper (2022)." http://www.caict.ac.cn/sytj/202209/P020220913580752910299.pdf.

Cheng, Qian. "Unveiling the True Face of 94 ChatGPT Concept Stocks." https://36kr.com/p/2124450621753476.

CMB International. "ChatGPT Cross-Industry Research Report: AIGC Development Fuels the Next Industrial Revolution." https://www.cmbi.com.hk/article/7840.html?lang=cn.

Dongwu Securities. "ChatGPT Special Report: Another Theme of 'Humanoid Robots.'" https://pdf.dfcfw.com/pdf/H3_AP202301211582259865_1.pdf?1674742365000.pdf.

Fu, Yao, Peng Hao, Tushar Khot, and Guo Zhijiang. "Deconstructing GPT-3.5: Tracing the Origins of Its Various Abilities." https://yaofu.notion.site/GPT-3-5-360081d91ec245f29029d37b54573756.

Forbes. "Exclusive Interview with OpenAI Founder: How Will ChatGPT and AGI Break Capitalism?" https://www.forbeschina.com/leadership/63140.

Guosheng Securities. "Overseas Tech Giants Follow Suit, Prospect of Commercializing Chat-GPT." https://www.fxbaogao.com/detail/3555914.

Huang, Leping, and Zhang Haoyi. "A Cold Analysis of the ChatGPT Craze." https://www.tiantianyanbao.com/article/19067.html.

Huaxi Computer Team. "ChatGPT Industry Research: Opening a New Era of AI." http://pg.jrj.com.cn/acc/Res/CN_RES/INDUS/2023/2/1/5f0f1400-18a4-4955-9bb4-37b8ce1e78cd.pdf.

Kahn, Jeremy, and Liu Jinlong. "Fortune Cover Story: How Was ChatGPT, the Global Sensation, Created?" https://finance.sina.com.cn/tech/it/2023-01-31/doc-imyeawtc2784632.shtml.

Lin, Zhijia. "China and the US Compete to Develop ChatGPT: Understanding the Latest Layouts of Over 20 Tech Giants." https://www.tmtpost.com/6402165.html.

NewSmart. "Microsoft Invests Another $10 Billion in OpenAI! Who Will Be the Biggest Winner in the AI Game Among Tech Giants?" https://36kr.com/p/2083113026351617.

Peng, Danni. "Why is ChatGPT So Powerful and Still Evolving?" http://www.stcn.com/article/detail/792655.html.

Qi, Jian, and Chen Yifan. "ChatGPT Triggers a New Round of Technology Arms Race." https://finance.sina.com.cn/tech/internet/2023-02-08/doc-imyexvzy6372104.shtml.

Ren, Xiaoning. "The Exploding Popularity of ChatGPT Will Trigger the Large-Scale Commercialization of AI." https://36kr.com/p/2115519307516036.

Synced. "Google Officially Releases Bard to Challenge ChatGPT, CEO Invites Testers Personally." https://finance.sina.com.cn/tech/csj/2023-02-07/doc-imyevtat6124555.shtml.

Tencent Research Institute. "AIGC Development Trend Report 2023: Embracing the Next Era of AI." http://www.cecc.org.cn/m/202302/570932.html.

Thompson, Ben. "AI and the Big Five." Stretcher by Ben Thompson. https://stratechery.com/2023/ai-and-the-big-five/.

Urban, Tim. "The AI Revolution: The Road to Superintelligence." https://waitbutwhy.com/2015/01/artificial-intelligence-revolution-1.html.

Zhang, Xinyi. "ChatGPT Goes Viral, NVIDIA Smiles." http://www.cena.com.cn/semi/20230213/118946.html.

Index

A

Afeyan, Noubar, 139

AI-generated content (AIGC), 5, 6, 11–16, 26, 27, 52, 60, 61, 63, 65–67, 71, 75, 76, 83, 84, 85, 86, 88, 97, 99–101, 103, 104, 108, 109, 110, 111, 112, 113, 126–29, 131, 132, 147, 169, 171, 176

Alibaba, x, 5, 38, 85, 87, 92–96, 100, 103, 107, 113, 115, 117, 149, 150, 170, 178

Alibaba Cloud, x, 5, 38, 85, 87, 92–96, 100, 103, 107, 113, 115, 117, 149, 150, 170, 178

AlphaFold, 59, 179

AlphaGo, 6, 23, 39, 59, 97, 159

AlphaTensor, 59

Amazon, 44, 62, 66–71, 73, 117, 158, 164

Amazon Web Services (AWS), 44, 67–69

Anthropic, 57

App Tracking Transparency (ATT), 65

B

Baidu, x, 5, 38, 85–89, 91, 92, 100, 103, 111, 113–15, 117, 122, 126, 170, 171

Bard, x, 36, 57, 58, 61, 62, 85, 126

Bing, 5, 12, 33, 45, 50–53, 58, 85, 100, 123–26

Buchheit, Paul, 58

ByteDance, 86, 100–103, 113

C

ChatGPT Plus, x, 49

Chinese Language Understanding Evaluation (CLUE), 97, 98

Cook, Tim, 61, 189

CUDA (Compute Unified Device Architecture), 72, 74

D

DALL·E, 12, 13, 14, 26, 45, 47–49, 60, 178

data2vec, 64, 65

DeepMind, 43, 44, 58, 59, 64, 179

E

ERNIE, x, 85–92, 111, 114, 126

F

FAIR (Facebook AI Research), 64

FBA (Fulfillment by Amazon), 69, 71

G

Galactica, 65

Gates, Bill, x, 6, 71, 193

Google, x, 4, 7, 8, 14, 34, 35, 36, 43–45, 49, 50, 52, 53, 55–62, 64, 67, 68, 73, 74, 76, 77, 78, 85, 86, 88, 93, 100, 114, 117, 121, 122, 125, 126, 169, 170

Google Brain, 7, 8, 35, 59

Google Cloud, 57, 60, 74

Gordon Moore, 76

GPT-2, 8, 9, 39, 47

Gurman, Mark, 63

H

Huang, Jensen, 71, 72

Huawei, 86, 103, 105, 114, 116, 117

Hunyuan, 97–99

I

Imagen, 60, 67, 170

ImageNet, 73

InstructGPT, 9, 10, 48, 113

J

JD.com (JD), 38, 85, 86, 95, 96, 104–7, 150

K

Karpathy, Andrej, 78

K-PLUG, 104, 107

Kunlun, 88, 112

L

LaMDA, 57, 60, 67, 88

LeCun, Yann, 63, 64

Lee Sedol, 6, 39, 59

Lessin, Sam, 66

Li, Robin, 87, 88, 89, 90, 92, 171

M

Ma, Jack, 94, 149

Meta, 63–66, 103, 117, 170

Microsoft, x, 5, 12, 13, 33, 36, 45, 46, 48–55, 57, 58, 61, 62, 67, 68, 69, 73, 74, 78, 81, 85, 86, 87, 92, 114, 117, 123, 125, 126, 128, 129, 130, 166, 170, 171

Mostaque, Emad, 60, 61, 68

Musk, Elon, x, 43, 77, 156

N

Nadella, Satya, 12, 45, 51, 125

Ng, Andrew, 87

NVIDIA, 65, 68, 69, 71–77, 166

O

OpenAI, ix, x, 1, 2, 3, 5, 6, 9, 10, 11, 27–35, 39, 41, 43–51, 55, 57, 61, 63, 64, 69, 74, 77–79, 83, 85, 86, 90, 108, 114, 117, 138, 142, 166, 170, 171, 172, 176, 178, 180

P

PageRank algorithm, 56

Pan Jianwei, 77

Pengcheng Panggu, 114

Pichai, Sundar, 57, 67, 126

R

Reality Lab, 64

S

SenseTime, 109–11

Sequoia Capital, 42, 83

Siri, 7, 11, 23, 62, 63

Stability AI, 13, 60, 61, 67, 68, 171, 172

Stable Diffusion, 11, 13, 14, 26, 61, 62, 67, 69, 74, 170, 171, 172

Sycamore, 76

T

Tangle Lake, 77

Tencent, x, 85, 86, 87, 97–100, 103, 113, 115, 117, 127, 135, 150, 170

TikTok, ix, 101–3, 117

Tongyi Qianwen, x

Transformer, 2, 7, 8, 34, 35, 36, 59, 60, 65, 91, 101, 102, 112, 169

Treebank, 8, 59

Twitter, 13, 41, 46, 77, 79–82

W

WeChat, 99, 100, 152, 171

Wozniak, Steve, 82

Y

Yanxi, 105–7

Yu, Kai, 87

Z

Zhou Jingren, 93, 94

ABOUT THE AUTHOR

Mr. Kevin Chen is a renowned science and technology writer and scholar, visiting scholar at Columbia University, a postdoctoral scholar at the University of Cambridge, and an invited course professor at Peking University. He has served as a special commentator and columnist for the *People's Daily*, CCTV, China Business Network, SINA, NetEase, and many other media outlets. He has published several monographs involving numerous domains, including finance, science and technology, real estate, medical treatment, and industrial design. He has currently taken up residence in Hong Kong.